Trap Level Spectroscopy in Amorphous Semiconductors

Trap Level Spectroscopy in Amorphous Semiconductors

Victor I. Mikla and Victor V. Mikla

ELSEVIER

AMSTERDAM • BOSTON • HEIDELBERG • LONDON • NEW YORK • OXFORD
PARIS • SAN DIEGO • SAN FRANCISCO • SINGAPORE • SYDNEY • TOKYO

Elsevier
32 Jamestown Road London NW1 7BY
30 Corporate Drive, Suite 400, Burlington, MA 01803, USA

First edition 2010

British Library Cataloguing-in-Publication Data
A catalogue record for this book is available from the British Library

Library of Congress Cataloging-in-Publication Data
A catalog record for this book is available from the Library of Congress

ISBN: 978-0-323-16503-7

For information on all Elsevier publications visit
our website at www.elsevierdirect.com

Typeset by MPS Limited, a Macmillan Company, Chennai, India
www.macmillansolutions.com

This book has been manufactured using Print On Demand technology. Each copy is produced to order and is limited to black ink. The online version of this book will show color figures where appropriate.

Transferred to Digital Printing in 2016

Contents

Acknowledgments

I would like to express my sincere thanks to Lisa Tickner (Elsevier), Lisa Jones (Elsevier), and Mani Prabakaran (MPS Limited) for their patience, continued interest, and helpful comments, which have made it possible for this book to reach completion.

Undoubtedly, without the inspiration of my wife, Ottilia, this material could never have become a book.

Victor I. Mikla
Uzhgorod
March 2010

1 Defect States Spectroscopy in Amorphous Semiconductors

Contents

1.1 Introduction

Amorphous semiconductors are characterized by properties that are absent in their crystalline counterparts. On the one hand, they are "unsuitable" objects both for experimentalists and from the theoretical point of view; on the other hand, they have widespread technical applications. Therefore, trap level spectroscopy in materials containing S, Se, and Te is necessary for further technical applications. Currently, there is no universal technique that probes the full spectrum of trapping levels in a mobility gap; experimentalists use several complementary methods. This book is devoted to techniques that probe states in the mobility gap and the results of their use.

The book is an encyclopedia in the sense that it gives "starting" information about a wide range of spectroscopic techniques in disordered materials. The author only provides brief, annotated descriptions of these techniques, their advantages, and experimental results that are typical for the objects considered.

Trap Level Spectroscopy in Amorphous Semiconductors. DOI: 10.1016/B978-0-12-384715-7.00001-2

1.2 General Principles

For thermally stimulated processes to occur, fulfillment of two general conditions is necessary:

a. The system considered must be removed from equilibrium and must exist in a state with factors present such that it does not allow reestablishment of equilibrium.
b. The system must have a contact with a temperature reservoir (the latter provides the thermal energy necessary to activate the relaxation process).

Of special interest for thermally stimulated relaxation (TSR) is how to remove the system from equilibrium and physical phenomena that can be measured (monitored) during the relaxation process. We restrict ourselves by considering the physical phenomena, although these can also take place in chemical and biological objects. Further, among physical process, we consider only those that involve redistribution of electronic charge carriers in semiconductors during the relaxation process.

1.2.1 Basic Types of Relaxation Techniques

Basic types of relaxation techniques used are:

a. *Isothermal relaxation*: In this case the perturbation is implemented at a constant temperature that is selected to obtain experimentally convenient relaxation times.
b. *Nonisothermal relaxation*: In this case the system is perturbed at temperatures sufficiently low to reestablish a new statistical equilibrium. In the following, the temperature is increased according to a well-controlled heating program $T(t)$, thus increasing the reaction rates, and the relaxation of the system can be monitored as a function of temperature and time.

The first technique is successfully used not only in the study of chemical reactions but also in electronic reaction kinetics in solids. It is necessary to note here the recently developed technique of deep-level transient spectroscopy (DLTS).

Typically, nonisothermal relaxation is effectively employed in the studies of thermally stimulated luminescence (TSL), conductivity (TSC), polarization, and depolarization.

1.2.2 The Principle of Detailed Balance and Classification of Trapping States

The rate at which a perturbed system reacts during the relaxation process has been described by an equation of the form

$$\frac{d[n_D]}{dt} = \alpha[n_A]^m[n_B]^n[n_C]^l \tag{1.1}$$

where n_D is the concentration of the product resulting from the reaction between the reactants n_A, n_B, and n_C, and α is the rate coefficient. It is necessary to note here that this is only a phenomenological description of measured rates and can be related to actual physical processes only in very simple cases.

Equation (1.1) applies only to systems that are removed from thermal equilibrium and where the reaction reasonably occurs in one direction.

The principle of detailed balance is a result of the microscopic reversibility of electron kinetics. A prerequisite for the establishment of thermal equilibrium requires that the forward and reverse rates are identical. For isothermal reactions, the equilibrium constant remains unchanged. The principle of detailed balance is of fundamental importance to establish helpful relations between reaction and equilibrium constants because both are at the initial thermal equilibrium; in addition, at the new equilibrium after the relaxation of the perturbation, the net forward and reverse reaction rates are zero.

Defect states in the forbidden (mobility) gap may act as electron or hole traps depending on their states of occupancy. The introduction of quasi-Fermi levels for trapped electron and holes allows the classification of trapping states as shallow or deep traps. The involvement of electron states in the reaction process decreases with increasing energy from the quasi-Fermi level for trapped electrons up to the edge of the conduction band, or with decreasing energy from the quasi-Fermi level for trapped holes down to the top of valence band. Free carriers falling into one of these trapping states will be reemitted with high probability back into the band from which they came from. These states are empty and are usually termed as shallow states. Evidently, the temperature of establishing the degree of occupancy will determine whether or not it acts as a shallow trap. Reasonably, we use such a definition of a shallow trapping state as a "statistical" one in contrast to the distinction between shallow and deep trapping states provided by physical arguments, namely, that shallow traps are characterized by a very small ionization energy that is of the order of the phonon energies. Deep traps are those whose ionization energy is many times that of the phonon and, consequently, capture of free carrier may involve multiphonon transitions. Reasonably, at temperatures above several degrees Kelvin, shallow trapping states are empty; in such a case, both definitions of shallow traps are identical. On the other hand, at elevated temperatures, deep trapping states may turn into "statistically" shallow traps.

For all TSR processes (with the corresponding redistribution of electrons and holes over the states in the gap) to be discussed, it seems useful to give a criterion permitting the classification of states under consideration as traps and recombination centers, respectively.

We have previously used the generic term "trap" for all states in the gap. However, a clear distinction is possible as to whether, at a given temperature and for a given occupation function, the probability for a carrier to be released from the defect state into one of the two allowed bands is larger than the probability for capture of a free carrier. This is done by introducing so-called demarcation levels, E_d^n, for trapped electrons and, E_d^p, for trapped holes. According to Rose [1], E_d^n is the energy level in a trap distribution at which the electron has equal probabilities of being thermally released into the conduction band or of recombining with a free hole.

It is now obvious that all states between E_d^n and E_v are recombination centers for electrons, while all states between E_d^p and E_c are recombination centers for holes. In first approximation, E_d^n is somewhat smaller than the quasi-Fermi level for free electrons and E_d^p is somewhat larger than quasi-Fermi level for free holes (Fig. 1.1).

and trapped carriers. It should be mentioned that none of the TSR techniques listed here is suitable to identify the microscopic structure and chemical nature of centers involved. To obtain such information, independent experiments are needed.

1.2.4 Direct Methods

The reaction or thermal escape rate can be monitored directly and can be determined by measuring the concentration of trapped carriers as a function of time and/or temperature. This is accomplished by using the material to be studied in a capacitor configuration (e.g., as a p–n junction, as a Schottky barrier or, in general, as thin film sandwiched between electrodes) and by recording the change in capacitance during the relaxation process. The capacitance change may be measured isothermally at one or several fixed temperatures, and the experimental techniques employed have become known as isothermal capacitance transients (ICAPTs) [3], Deep Level Transient Spectroscopy (DLTS) [4]. These methods differ in their experimental sophistication, convenience in use, and sensitivity.

Nonisothermal capacitance methods employ heating programs $T(t)$ and are known as thermally stimulated capacitance (TSCAP) [5].

All other experimental TSR techniques used in trap level spectroscopy in semiconductors (insulators) are indirect methods for the determination of trapping parameters. The techniques involve the measurement of phenomena that are due to charge carriers emitted after thermal stimulation from the traps.

1.2.5 Indirect Methods

A carrier thermally released from the trap into the transport band may be either retrapped by the same species of traps or a different one and, under the influence of an electric field, may contribute to an externally measurable current. It may either be swept out of the region being probed or recombined with a recombination center. Some of the electrons may even overcome the work function barrier and leave the solid. The traffic of these carriers from traps to the recombination centers or out of the material can be monitored at various stages, and thus, information on the thermal emission rates can be obtained indirectly.

During the TSR process, the concentration of holes and electrons is determined by the balance between thermal emission and recapture by traps and capture by recombination centers. In principle, integration of corresponding equations yields $n_C(t,T)$ and $p(t,T)$ for both isothermal current transients (ICTs) or during irreversible thermal scans. Obviously, the trapping parameters listed together with the capture rates of carriers in recombination centers determine these concentrations. Measurement of the current density $J = \exp(\mu_n n_C + \mu_p p)$ will provide trap-spectroscopic information. The experimental techniques employed in an attempt to perform trap level spectroscopy on this basis are known as Isothermal Current Transients (ICTs) [6], TSC [7].

An additional method is thermally stimulated capacitor discharge (TSCD) [8]. It involves filling the traps at some "high" temperature (e.g., room temperature) by the application of the high field and subsequent cooling to a lower temperature with the field applied. Thereafter, the field is removed and the sample heated in the usual

Equation (1.1) applies only to systems that are removed from thermal equilibrium and where the reaction reasonably occurs in one direction.

The principle of detailed balance is a result of the microscopic reversibility of electron kinetics. A prerequisite for the establishment of thermal equilibrium requires that the forward and reverse rates are identical. For isothermal reactions, the equilibrium constant remains unchanged. The principle of detailed balance is of fundamental importance to establish helpful relations between reaction and equilibrium constants because both are at the initial thermal equilibrium; in addition, at the new equilibrium after the relaxation of the perturbation, the net forward and reverse reaction rates are zero.

Defect states in the forbidden (mobility) gap may act as electron or hole traps depending on their states of occupancy. The introduction of quasi-Fermi levels for trapped electron and holes allows the classification of trapping states as shallow or deep traps. The involvement of electron states in the reaction process decreases with increasing energy from the quasi-Fermi level for trapped electrons up to the edge of the conduction band, or with decreasing energy from the quasi-Fermi level for trapped holes down to the top of valence band. Free carriers falling into one of these trapping states will be reemitted with high probability back into the band from which they came from. These states are empty and are usually termed as shallow states. Evidently, the temperature of establishing the degree of occupancy will determine whether or not it acts as a shallow trap. Reasonably, we use such a definition of a shallow trapping state as a "statistical" one in contrast to the distinction between shallow and deep trapping states provided by physical arguments, namely, that shallow traps are characterized by a very small ionization energy that is of the order of the phonon energies. Deep traps are those whose ionization energy is many times that of the phonon and, consequently, capture of free carrier may involve multiphonon transitions. Reasonably, at temperatures above several degrees Kelvin, shallow trapping states are empty; in such a case, both definitions of shallow traps are identical. On the other hand, at elevated temperatures, deep trapping states may turn into "statistically" shallow traps.

For all TSR processes (with the corresponding redistribution of electrons and holes over the states in the gap) to be discussed, it seems useful to give a criterion permitting the classification of states under consideration as traps and recombination centers, respectively.

We have previously used the generic term "trap" for all states in the gap. However, a clear distinction is possible as to whether, at a given temperature and for a given occupation function, the probability for a carrier to be released from the defect state into one of the two allowed bands is larger than the probability for capture of a free carrier. This is done by introducing so-called demarcation levels, $E_{\mathrm{d}}^{\mathrm{n}}$, for trapped electrons and, $E_{\mathrm{d}}^{\mathrm{p}}$, for trapped holes. According to Rose [1], $E_{\mathrm{d}}^{\mathrm{n}}$ is the energy level in a trap distribution at which the electron has equal probabilities of being thermally released into the conduction band or of recombining with a free hole.

It is now obvious that all states between $E_{\mathrm{d}}^{\mathrm{n}}$ and E_{v} are recombination centers for electrons, while all states between $E_{\mathrm{d}}^{\mathrm{p}}$ and E_{c} are recombination centers for holes. In first approximation, $E_{\mathrm{d}}^{\mathrm{n}}$ is somewhat smaller than the quasi-Fermi level for free electrons and $E_{\mathrm{d}}^{\mathrm{p}}$ is somewhat larger than quasi-Fermi level for free holes (Fig. 1.1).

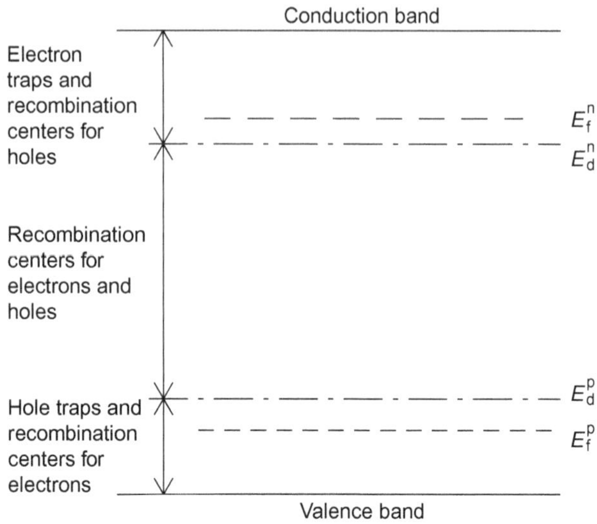

Figure 1.1 Demarcation levels E_d^n and E_d^p of semiconductor with an arbitrary distribution of traps.

It is necessary to make a few remarks about relaxation of the system after the external perturbation is removed.

In fact, after removal of the source of external perturbation, the system is left in a nonequilibrium state, the relaxation of which may be considered to occur in two phases:

1. The excess free carrier density decays (via recombination) with those recombination centers that have the largest cross section.
2. Further relaxation toward thermal equilibrium proceeds via thermal release of carriers from traps into the bands and subsequent recombination with recombination centers.

The rate limiting step of this second phase is the thermal release of trapped carriers. The probability of this process can be enhanced by increasing the temperature. During this process, the quasi-Fermi levels and demarcation levels move toward the equilibrium Fermi level. It is necessary to make several assumptions to describe the relaxation kinetics:

1. Band–band transitions are relatively unlikely to occur as compared to free carrier recombination with the corresponding (recombination) centers.
2. Transitions between trapping states for electrons above E_d^n and for holes below E_d^p are nonradiative transitions and involve the emission or absorption of phonons only.
3. Transitions between recombination centers and the corresponding bands may be radiative transitions involving the emission or absorption of photons.
4. The system (i.e., the nonequilibrium distribution of electrons and holes over available energy states) is in thermal contact with black body radiation of density $g(E)$ and phonons of energy density $g_p(E)$. The mentioned densities are the equilibrium densities at the temperature T of the solid.

Now we can write

$$\frac{dn_c}{dt} = -\int_{E_v}^{E_d^n} [a(E) + b(E)g(E)]Vn_c N(E)[1 - f(E)]dE$$
$$+ \int_{E_v}^{E_d^n} [b(E)g(E)Vn_c N(E)f(E)]dE$$
$$+ \int_{E_d^n}^{E_\infty} [a(E)N(E)f(E)]dE$$
$$- \int_{E_d^n}^{E_\infty} \beta(E)n_c N(E)[1 - f(E)]dE \qquad (1.2)$$

Here $a(E)$ and $b(E)$ are the Einstein coefficients for spontaneous and induced transition probabilities, respectively; V the volume of the solid; $a = \langle v_n \rangle S_n \exp[-(E_c - E_v)/kT]$; and $\beta = \langle v_n \rangle S_n$. Note that the capture cross section S_n is now that of the single or multiphonon transition.

It is also necessary to note that the success of TSR techniques to obtain information on trapping states in the gap depends on whether or not the experiment can be performed under conditions that justify equation (1.2) to be reduced to simple expressions for the kinetic process. Usually, the kinetic theory of TSR phenomena in bulk semiconductors—such as thermoluminescence, thermally stimulated current, polarization, and depolarization—has been interpreted by simple kinetic equations that were arrived at for reasons of mathematical simplicity only and that had no justified physical basis. The hope was to determine the most important parameters of traps—namely, the activation energies, thermal release probabilities, and capture cross section—by fitting experimental curves to those oversimplified kinetic descriptions. The success of such an approach seems to be only marginal. This situation changed after it was realized that TSR experiments can indeed be performed under conditions that justify the use of simple theoretical approaches for the determination of trapping parameters:

a. Instead of dealing with a distribution of traps only a limited number of trap levels is present.
b. The experimental conditions must eliminate certain transitions that complicate the kinetic equations, e.g., retrapping may be neglected under high-field conditions in thermally stimulated current experiments.

1.2.3 Trap Level Spectroscopy: Experimental Methods

All TSRs involve the release of trapped charge carriers into either the conduction band or valence band and their subsequent capture by recombination centers and recapture by other traps (retrapping). Their experimental investigation is undertaken with the goal of determining the characteristic properties (parameters) of traps: capture cross sections, thermal escape rates, activation energies, concentration of traps, and trapped carriers. It should be mentioned that none of the TSR techniques listed here is suitable to identify the microscopic structure and chemical nature of centers involved. To obtain such information, independent experiments are needed.

1.2.4 Direct Methods

The reaction or thermal escape rate can be monitored directly and can be determined by measuring the concentration of trapped carriers as a function of time and/or temperature. This is accomplished by using the material to be studied in a capacitor configuration (e.g., as a p–n junction, as a Schottky barrier or, in general, as thin film sandwiched between electrodes) and by recording the change in capacitance during the relaxation process. The capacitance change may be measured isothermally at one or several fixed temperatures, and the experimental techniques employed have become known as isothermal capacitance transients (ICAPTs) [3], Deep Level Transient Spectroscopy (DLTS) [4]. These methods differ in their experimental sophistication, convenience in use, and sensitivity.

Nonisothermal capacitance methods employ heating programs $T(t)$ and are known as thermally stimulated capacitance (TSCAP) [5].

All other experimental TSR techniques used in trap level spectroscopy in semiconductors (insulators) are indirect methods for the determination of trapping parameters. The techniques involve the measurement of phenomena that are due to charge carriers emitted after thermal stimulation from the traps.

1.2.5 Indirect Methods

A carrier thermally released from the trap into the transport band may be either retrapped by the same species of traps or a different one and, under the influence of an electric field, may contribute to an externally measurable current. It may either be swept out of the region being probed or recombined with a recombination center. Some of the electrons may even overcome the work function barrier and leave the solid. The traffic of these carriers from traps to the recombination centers or out of the material can be monitored at various stages, and thus, information on the thermal emission rates can be obtained indirectly.

During the TSR process, the concentration of holes and electrons is determined by the balance between thermal emission and recapture by traps and capture by recombination centers. In principle, integration of corresponding equations yields $n_C(t,T)$ and $p(t,T)$ for both isothermal current transients (ICTs) or during irreversible thermal scans. Obviously, the trapping parameters listed together with the capture rates of carriers in recombination centers determine these concentrations. Measurement of the current density $J = \exp(\mu_n n_C + \mu_p p)$ will provide trap-spectroscopic information. The experimental techniques employed in an attempt to perform trap level spectroscopy on this basis are known as Isothermal Current Transients (ICTs) [6], TSC [7].

An additional method is thermally stimulated capacitor discharge (TSCD) [8]. It involves filling the traps at some "high" temperature (e.g., room temperature) by the application of the high field and subsequent cooling to a lower temperature with the field applied. Thereafter, the field is removed and the sample heated in the usual manner. The current, measured during heating, consists of two components: (a) the dielectric relaxation current and (b) the current due to carriers thermally released from the trap into the two upper bands. For TSCD to be utilized as a trap-spectroscopic tool, one has to subtract component (a) from the total current. Component (b) is, of course, closely related to TSC.

One of the most commonly employed techniques is TSL, which monitors photons as a function of temperature during the thermal scan. These photons are the result of radiative transitions (luminescence) of free carriers, previously released from the traps, to recombination centers.

Thermally stimulated dielectric relaxation of a solid in a polarized (electret) state usually was selected for the following reasons:

1. This experimental method, as well as the formal kinetics of the process, is closely related to trap level spectroscopy by thermally stimulated release of trap charge carriers.
2. Thermally stimulated discharging of electrets provides spectroscopic information similar to trap level spectroscopy; most importantly, their density and the activation energy required for the relaxation process to proceed.

The polarized state (nonequilibrium steady state) is created by applying a DC voltage at an elevated temperature and by subsequent cooling of the solid to a temperature that is sufficiently low that rapid relaxation is prevented. The next step of the experimental procedure is to remove the DC bias. The currents that can be measured during either isothermal or nonisothermal relaxation back to thermal equilibrium are used to monitor the relaxation processes involved.

A detailed discussion of the statistical thermodynamic aspects of thermally stimulated dielectric relaxation is not provided here. It should suffice to state that kinetics of most of the processes are again complicated and that the phenomenological kinetic theories used to described thermally stimulated currents make use of assumptions that, being necessary to simplify the formalism, may not always be justified. Just as in the general case, TSL and TSC, the spectroscopic information may in principle be available from the measurement of thermally stimulated depolarization current (TSDC). However, it is frequently impossible to extract it unambiguously from such experiments.

1.2.6 TSL and TSC: Retrospective Glance

Observations of TSL have been reported as early as the seventeenth century, but Urbach [9] is generally credited with proposing it as a potentially useful experimental technique for trap level spectroscopy. However, only after the publication of the work of Randall and Wilkins [10] in 1945 did TSL receive much attention. First measurements of TSC and TSL were performed by Bube [11].

Both TSL and TSC are observed in a broad variety of materials, e.g., fish scale bones; dental enamel; plastics; and amorphous, polycrystalline, and crystalline semiconductors, to name a few.

The occurrence of TSL and TSC during a thermal scan of a previously excited material is probably the most direct evidence we have for the existence of electronic trap levels in these materials. A TSL/TSC spectrum usually consists of a number of more or less resolved peaks in luminescence intensity or electric conductivity versus temperature, which, in most cases, may be attributed to species of traps. Its appearance is a direct representation of the fact that

1. the escape probability of trapped carriers is a sharply increasing function with temperature;
2. the supply of trapped carriers is limited to start with and decreases with their ongoing thermal release from traps.

Because the escape probability of carriers from trapping sites is proportional to $\exp(-E/kT)$, the location of a glow peak on the temperature scale provides encoded information on the value of thermal activation energy E. Hence, a glow curve represents a spectrum of energies that are required to free carriers from the various species of traps in the material.

The procedure used to decode the glow spectrum and retrieve the desired trap-spectroscopic data appear obvious and straightforward—a measured curve is analyzed to obtain characteristics such as location of the peak on the temperature scale, its width, initial rise, and so forth. These data are then utilized to determine trapping parameter via an appropriate model for the reaction kinetic processes that occur during the temperature scan. However, exact knowledge of the proper kinetics is mandatory for this analysis to yield quantitative values.

The most simple of reaction kinetics that actually yield TSL and TSC peaks is the so-called single trap model [9] described later. The field of TSL and TSC developed along two lines. The first one merely made use of the capability of deep traps in certain insulating materials to store a charge at or below room temperature for a long time—sometimes thousands of years—without being overly concerned with the mechanism of this information storage and its eventual retrieval in a thermal scan in the form of thermoluminescence or TSC. One of the approaches concentrated on quantitative trap level spectroscopy and employing curve-fitting techniques on the basis of a single trap model and on efforts to completely understand the detailed features of TSL and TSC curves calculated within the framework of this model. A poor fit of experimental and computed glow curves was rarely related to inadequacy of the single trap model, but rather to the fact that only approximate solutions of its rate equations were available. Exact solutions of these differential equations were possible only after development of powerful numerical computation techniques.

It then became immediately evident that it is extremely difficult to correlate theory and experiment with any degree of confidence by curve fitting alone. Any measured and well-resolved glow peak that may reasonably be expected to be a result of a single type of trap can be fitted with a solution of the single trap model by approximately adjusting several of a set of many model parameters. Unfortunately, such a fit is not unique, because a number of different simple model descriptions are conceivable in addition to a single trap model. Very little effort has been extended to investigate them.

The origin of this lack of uniqueness has been traced to the fact that both TSC and thermoluminescence are only indirect trap-spectroscopic methods. In contrast to TSCAP techniques, the thermal release from traps or the capture of charge carriers in traps is not measured directly.

The full potential of TSL and TSC as quantitative spectroscopic tools has not yet been realized. In their present state of development, they give a quick and very sensitive survey of the number and relative concentration of traps—without being able to discriminate between hole and electron traps. Only estimates of activation energy are possible.

1.2.7 Defect States in Semiconductors and Insulators

The primary objects of investigation in both TSL and TSC experiments are nonradiative transitions between the ground level and trap and the conduction band and/or valence band. In a case when two different impurities—a donor and an acceptor—are close enough that their functions overlap, tunneling processes may take place from an excited level of the donor to the acceptor. In this case, the excited level can be occupied from the ground state via a nonradiative transition. In a case when recombination traffic proceeds via tunneling between donor–acceptor pairs, no TSC will be observed.

In general, various types of traps and recombination centers may be present, and their involvement in the reaction kinetic process will greatly change with temperature. The temperature range in which a specific range dominates must, therefore, be determined. This is most conveniently achieved with the aid of nonisothermal temperature scans, during which TSL and TSC are monitored. Of course, the microscopic physical and chemical nature of traps cannot be determined with these methods.

All trap-spectroscopic techniques that are based on thermal transport properties have in common that the interpretation of empirical data is often ambiguous because it requires knowledge of the underlying reaction kinetic model. Consequently, a large number of published trapping parameters—with the possible exception of thermal ionization energies in semiconductors—are uncertain. Data obtained with TSC and TSL techniques, particularly when applied to photoconductors and insulators, are no exceptions.

The temperature dependence of various thermal transport phenomena can be measured isothermally at a number of different temperatures where the sample is in thermal equilibrium, in steady-state equilibrium, or decays after pulsed excitation in a transient fashion. In contrast, TSL and TSC experiments are nonisothermal and observed only during a programmed change in a sample temperature.

The very large number of publications on characterizing traps precludes a thorough review in this chapter. Thus, only monographs and classic papers in the fields are referenced.

If one were to judge the importance of TSL and TSC relative to other thermal transport methods for trap characterization by number of recently published papers, these techniques appear to far outweigh all others combined, even though about one-third of the publications are concerned with applications in dosimetry and related TSL (TSC) instrumentation. However, the large number of articles on TSL and TSC does not necessarily indicate any advantage in their usefulness as trap-spectroscopy tools over the other methods. What can safely be concluded is that nonisothermal TSR is still, at the present time, a very active field of research.

Let us now briefly consider the various steps in a typical TSL or TSC experiment in wide-band gap material. We choose electromagnetic radiation as a means of excitation. The interaction of this radiation with the solids leads to a number of electronic phenomena, many of which are not clearly understood. They include the production of new defects as well as filling of trap levels with electron and holes:

1. High-energy electromagnetic radiation (X, γ, UV) produces hot electrons in the crystal that may multiply by impact ionization and subsequently quickly thermalize so that one is left with free carriers and excitons. A fraction of the incoming energy gives rise to radiation damage defects.
2. The excess free carriers (and excitons) do not represent stable excited states of the solids. A fraction of them recombine directly after thermalization either radiatively or by multiphonon emission. In most materials, nonradiative transitions to defect states in the gap are the dominant mode of decay. The lifetime of free carriers $T = 1/\sigma v S$ is determined by the density σ of recombination centers, their thermal velocity v, and the capture cross section S, and may span $10\text{–}10^{-14}$ s. Electrons, captured by states above the demarcation level, and holes, captured by states below the hole demarcation level, may get trapped. The condition for trapping is given when the occupied electron trap has a very small cross section for recombining with a free hole. The trapping process has, until recently, not been well understood.
3. After the decay of the excess free carriers due to recombination and trapping transitions, the solid is in the so-called excited state, which is characterized by the perturbation of the statistical equilibrium. The concentration of the remaining free carriers is now determined by the balance between thermal emission of carriers from the traps, retrapping transitions, and capture by recombination centers.

If the excitation occurred at a low temperature such that the thermal emission rate of carriers from traps is very small, the perturbed equilibrium will exist for a long time and only upon an appropriate increase of the sample temperature can the relaxation process proceed at a rate that permits one to monitor it by measuring the conductivity $\sigma(T) = \exp(n_C\mu_n + p\mu_p)$ of the sample (TSC) or the luminescence (TSL) emitted by radiative recombination of carriers thermally released from the traps.

The principal goal of TSC trap level spectroscopy is to experimentally determine, by comparison of model glow curve with measured ones, the characteristic parameters that govern the nonisothermal relaxation kinetics of the solid.

1.2.8 Field-Induced Thermally Stimulated Currents

By heating a polarized sample up to or above the polarization temperature, the release of charges is gradually sped up and, when the half-life of this process becomes comparable with the timescale of experiment (the latter is determined by the heating rate), discharge becomes measurable and gives rise in the external circuit to a current that first increases with increasing temperature and then decays when the supply of charges is depleted (Fig. 1.2).

The method of field-induced thermally stimulated current (FITSC) consists of measuring—according to a definite (usually linear) heating scheme—the currents generated by the buildup and release of a polarized state in a high-resistivity solid

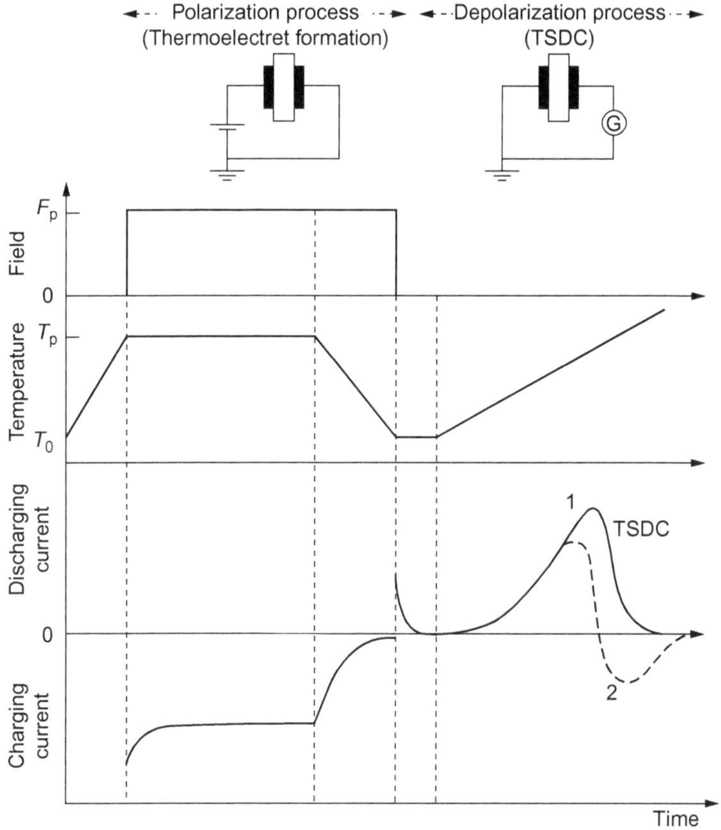

Figure 1.2 Thermoelectret formation and principle of the TSDC method: (1) heterocharge only present and (2) coexistence of hetero- and homocharges [2].

sandwiched between two electrodes. The standard experimental procedure involves the following steps:

1. the application of a DC bias V_p at a starting temperature T_p;
2. cooling under this temperature to some lower temperature T_0;
3. changing the bias at T_0 to another value V_d;
4. heating at a constant rate while maintaining the new bias;
5. recording the current as a function of temperature.

If the bias is zero, current peaks are observed during the thermal activation transition from the polarized state to the equilibrium state. This technique is known in the literature as TSDC.

The FITSC method is a general method of investigating the electrical properties of high-resistivity solids via the study of thermal relaxation effects and is an attractive alternative to conventional bridge methods, current–voltage temperature measurements, and so forth.

General aspects of the TSDC theory are discussed in an excellent review [2]. The main emphasis in this text is on experimental use of the TSDC method for amorphous chalcogenide semiconductors and on corresponding results for these intriguing objects.

1.2.9 Remarks on TSC/TSL

Approximate Numerical Solutions for TSC

Let us consider a solid that contains only one type of electron trap of volume density N at the discrete level E_t and a set of occupied deeper electron traps of density M. The experiment is performed in a temperature range in which traps N empty but traps M remain "thermally disconnected" and act only as an untapped reservoir of trapped electrons. The density of recombination centers is unspecified, but a density f of them is empty. At $T = T_0$, a concentration of f_0 empty recombination centers exists due to excitation. At these conditions, charge neutrality of the sample is of the form

$$f = n_C + n + M$$

Here, n_C is the density of free electrons and n that of occupied traps of type N. Now we can write

$$\frac{dn_C}{dt} = \alpha n - \beta n_C(N - n) - \gamma n_C(n_C + n + M)$$

We denote the capture coefficient for electron traps by β and that the recombination by γ. One can obtain

$$\frac{df}{dt} = \frac{dn_C}{dt} + \frac{dn}{dt} = -\gamma n_C(n_C + n + M)$$

as the following step. Numerical solutions of the preceding equations have been obtained (see references cited in [2]). Approximate analytic solutions have a long history and are possible if

$$n_C << n \quad \text{and} \quad \frac{dn_C}{dt} << \frac{dn}{dt}$$

yielding

$$-\frac{dn}{dt} = n(M + n)\alpha[(1 - R)n + M + RN]$$

$$n_C = \frac{\alpha n}{\gamma[(1 - R)n + M + RN]}$$

where $R = \beta/\gamma$ is the retrapping coefficient.

Analytic solutions for $\sigma(T)$ were reported by Simmons and Taylor [12] for the case that retrapping can be neglected in a thin sample at high electric fields. They considered the presence of several trap levels of density N_i and demonstrated the superposition of the individual glow peaks when the thermal ionization energies of these levels are very close to each other (Fig. 1.3).

Experimental Details

The main attraction of TSC and TSL as experimental methods for the study of trapping levels in high-resistivity solids was, for many years, their apparent simplicity. The excited sample merely had to be placed onto a heater in front of an optical detector or, after attachment of two metallic contacts for voltage biasing, connected to a sensitive current meter. Work at low initial temperatures required the experiment to be performed inside a vacuum chamber in an inert gas atmosphere.

As the field evolved, more complicated heating programs were found to be advantageous in certain situations. At the same time, the basic experimental arrangement was simple. In general, trap level spectroscopy by means of TSC and TSL requires the use of additional measurement techniques. Undoubtedly, which of these should be selected depends on the individual material and type of defect under study. As amorphous chalcogenide semiconductors were considered, time-of-flight (TOF) and xerographic spectroscopy seem to be the most informative techniques. Regarding TSC and TSL, it should be noted that these techniques have been used successfully in conjunction with independent techniques to shed light on the detailed properties of trap levels.

For most experiments on nonisothermal TSR, simple cooling of the sample to the desired initial temperature and a linear increase in T after excitation are sufficient to obtain TSC and TSL glow curves. Some techniques require more elaborate heating cycles, the details of which depend on the relaxation mechanism under study and on whether it is necessary to discriminate between simultaneously occurring processes, e.g., thermally stimulated depolarization and thermally stimulated conductivity (see Chapter 2).

It is important to note here that heating programs can be designed to overcome (at least partially) one of the major problems in the measurement of TSR—the occurrence of unresolved and/or overlapping peaks. The important role played by the heating rate in the amplitude as well as the position of the peak is obvious: when the heating rate increases, the initial polarization has to be released in a shorter time while the dielectric responds less quickly. Thus, the peak increases and shifts to a higher temperature. Another important consequence of a variation in a heating rate will be to change the resolution of the TSDC spectrum. When several relaxation peaks overlap each other, a decrease in heating rate should increase the resolving power. In practice, however, the use of large differences in heating rate is usually not possible (owing to probable

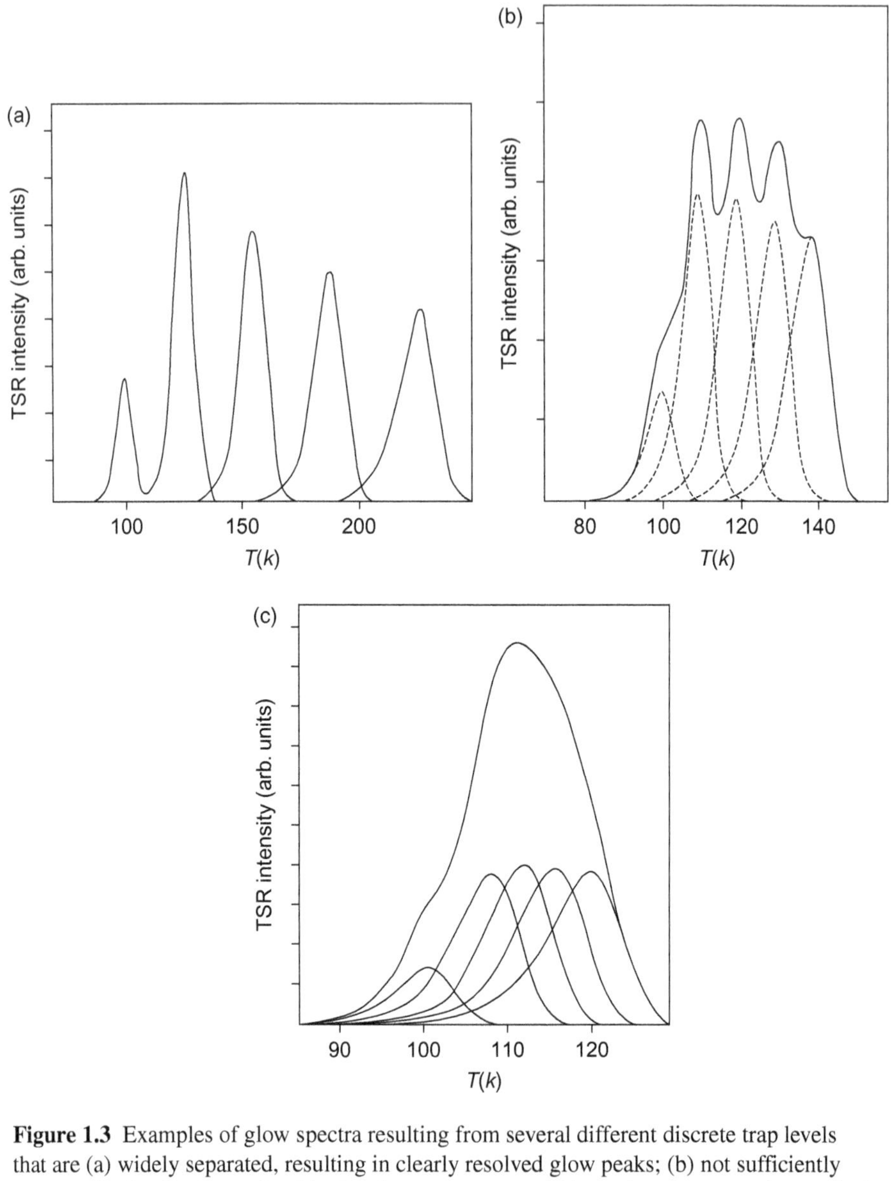

Figure 1.3 Examples of glow spectra resulting from several different discrete trap levels that are (a) widely separated, resulting in clearly resolved glow peaks; (b) not sufficiently separated to be clearly resolved in the glow spectrum; and (c) overlapping to produce a single, unresolved glow peak [12].

temperature lags), and the relative shifts and shape variations of neighboring peaks will thus be small. Furthermore, a decrease in heating rate will simultaneously involve a decrease in current intensity and thus in the signal-to-noise ratio. The choice of a heating rate for a given experiment will, therefore, rarely be determined by resolution considerations alone, but rather by a compromise taking into account the signal intensity and the possibility of temperature gradients in the sample.

Several experimental procedures have been used to overcome the problem of overlapping peaks: the decayed glow curve technique, thermal cleaning, and fractional emptying (Fig. 1.4). The first consists of selecting an initial temperature T_0 that is high enough that it does not fill those traps during excitation that produce the low-temperature peak (Fig. 1.4(a)). Thermal cleaning achieves this same result by preheating the sample, excited at T_0 to T_{01} (Fig. 1.4(b)).

Fractional emptying consists of measuring the initial rise of glow peaks (from which one can determine the thermal activation energy) during successive ramps of gradually increasing upper temperature limits. A histogram of the slopes, obtained from a semilogarithmic plot of the glow curve intensity versus $1/T$, reveals the spectrum of the activation energies involved in the unresolved glow curve. If these temperature ramps include cooling at the same rate as used during the heating ramp (Fig. 1.4(d)), the measured relaxation phenomenon will be reversible at certain temperature zones.

The rate q in a linear heating program $dT = q\,dt$ should be carefully considered. Usually, a compromise between fast heating (for signal-to-noise ratio to be improved) and uniform heating of the sample is chosen. The temperature difference between the back (heated) and front surface of a flat sample for a given heating rate q may be estimated from

$$\Delta T = \frac{qd^2c\xi}{k}$$

Here, c is the specific heat, k the thermal conductivity, ξ the sample density, and d its thickness. For a typical ionic material of $d = 1\,\text{mm}$, ΔT may reach more than 1 K with $q = 1\,\text{K/s}$. For amorphous chalcogenide semiconductors, the temperature region $-77 < T < 450\,\text{K}$ has been used most often. In this region, thermal activation energies in the range from 0.1 to 1.0 eV may be studied. The sample holder is cooled by liquid nitrogen and is inside a vacuum chamber to prevent condensation of moisture or electrical leakage. In a vacuum, the sample holder is usually a copper tip in contact with an external reservoir of liquid N_2. The sample is either clamped directly or attached to it with the aid of a contact agent of high thermal conductivity. If electrical isolation is required, thin platelets of Teflon, mica, and beryllium oxide have been used.

Standard optical metal or metal–glass cryostats are commonly used in low-temperature studies of TSL and TSC. Less expensive facilities for liquid N_2 temperatures have been designed, and closed cycle cryotips may be employed as well. TSR requires dynamic temperature cycles. Therefore, it is generally advantageous to design the sample holder-dewar arrangement as small as possible to enable quick turnaround times.

The heaters commonly consist of a resistive wire or a coaxial heating cable with an insulated heating wire inside bendable metal tubing. These are usually powered by DC supplies. Electrical leads and feedthroughs are designed for minimal leakage currents and stray capacitances.

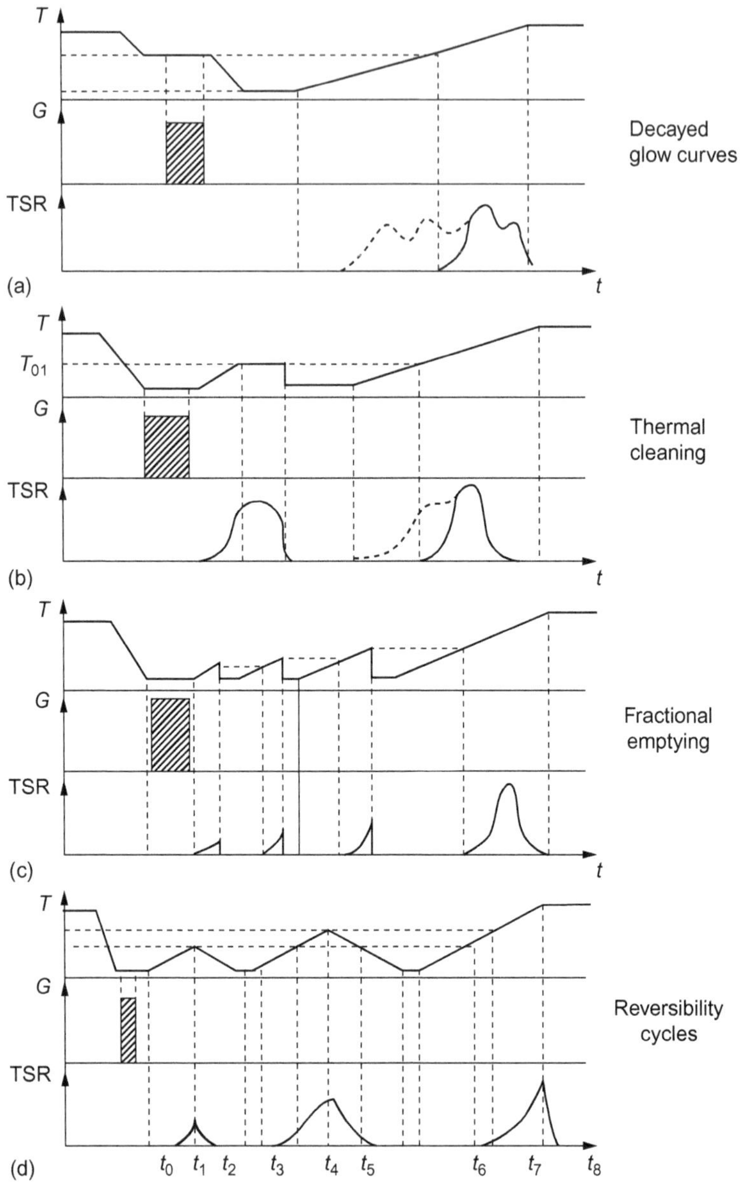

Figure 1.4 (a–d) Excitation (G) and heating cycles employed in TSR experiments. To remove disturbing glow peaks at low-temperature side of the peak under study, the sample is excited at a temperature chosen high enough to not fill traps responsible for the low-temperature peaks (a). The sample may also be excited at the low temperature but preheated prior to an appropriate higher temperature to measure the TSR peak under investigation (b). The technique of fractional emptying (c) is based on the existence of temperature zones in which TSR process is approximately reversible (d) [2].

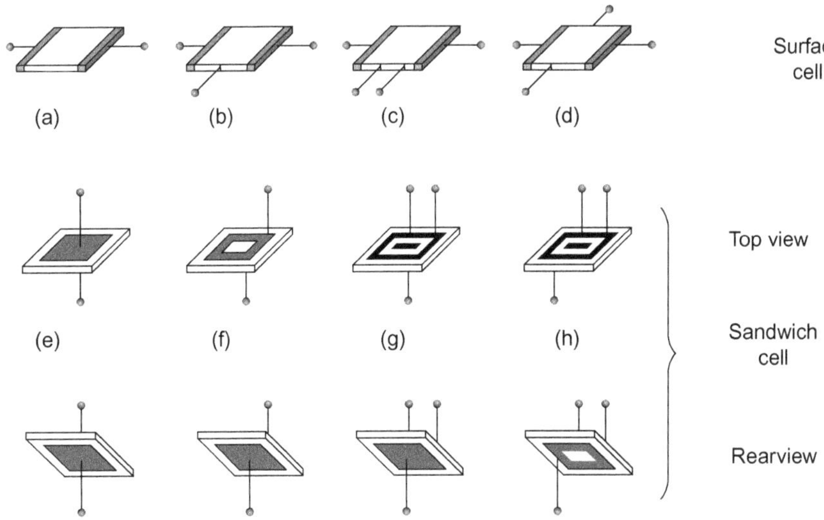

Figure 1.5 (a and b) Typical electrode configurations used for the measurements of thermally stimulated currents. The sample surface cell (a) may be augmented by additional contacts (b)–(d) to monitor potential distribution along the sample. The sandwich cell (e)–(h) is ideally suited for the use of guard rings of either rectangular or circular shapes.

For excitation in the visible range of the electromagnetic spectrum, lasers and gas discharge lamps, together with filters or monochromators, have been used.

TSC experiments are analyzed assuming the sample behaves "ohmic," i.e., the contacts do not introduce an inhomogeneity in the distribution of the electric field or carrier density and a uniform bulk density of carriers extends through the entire sample. Contact barriers are neglected.

Contact materials and configurations used in TSC experiments vary widely and depend on the particular application. Metal electrodes can be attached to the sample by evaporation or by application of conductive pastes (silver paint or epoxy) or metal-organic compounds. Typical contact configurations are shown in Fig. 1.5.

Biasing the sample is, in general, a simple matter. Stable DC supplies (floating or with one grounded outlet) or dry cell batteries are convenient to use because the currents through the sample are, in almost all cases, extremely small and require the use of DC picoampermeters with either linear or logarithmic output signal. Lock-in amplifiers can be considered whenever the signal can be modulated. It is necessary to note that the response and decay times in high-resistivity materials can be large. Therefore, the modulation frequencies have to be quite low.

In order to eliminate the uncertainties associated with transient current and field inhomogeneities along the sample during TSC experiments, it has been suggested that experiments be performed under constant-current conditions and compared with the usual constant voltage results. Constant-current sources for this type of test are commercially available.

The simplest way to obtain a record of TSC is a multiple-pen chart recorder that displays as a function of time $\sigma(T)$ and temperature. X–Y recorders are convenient for this purpose. They may serve to directly plot $\sigma(T)$ versus T. If initial rise techniques are used for the measurement of thermal ionization energies, a ln $[\sigma(T)]$ versus $1/T$ plot is suitable.

Estimation of Trap Parameters

In the experimental practice of TSDC, overlapping discharge processes are frequently present in most regions of the spectra. Therefore, the precise determination of the corresponding parameters requires previous separation of the specific processes involved. The TSDC method allows the easy and quick resolution of the overall spectrum. There are two ways in which this can be done.

The first, common to all nonisothermal techniques, was developed and applied to TSC and TSDC by Greswell and Perlman. Its principle is illustrated in Fig. 1.6.

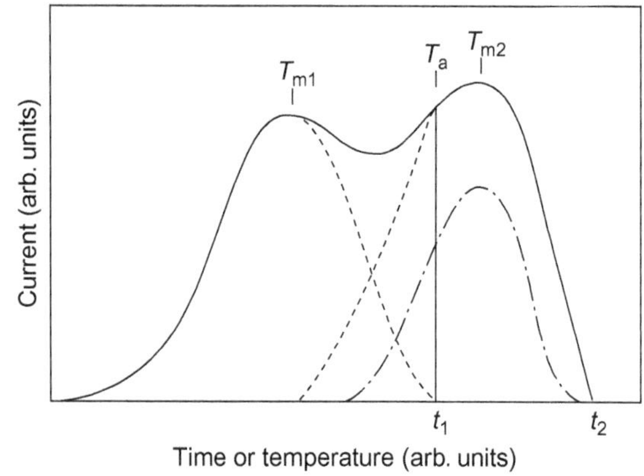

Figure 1.6 Principle of the peak-cleaning technique.

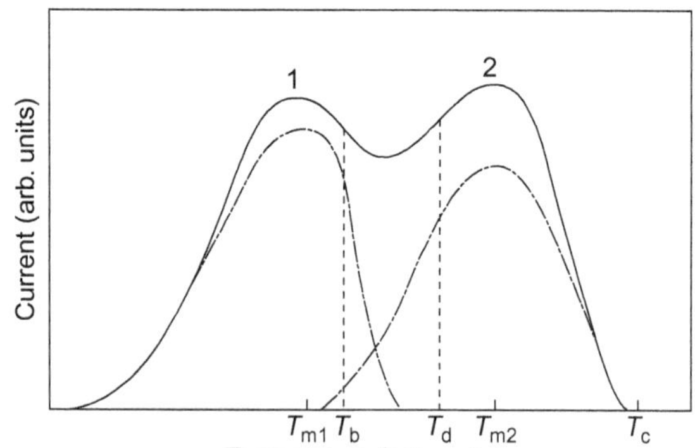

Figure 1.7 Principle of the peak-cleaning technique of Bucci et al. The dashed curves represent the TSDC peaks isolated during independent experiments by polarizing the sample from T_b to the lowest available temperature (peak 1) and from T_c to T_d (peak 2).

Let us suppose that we have two peaks, 1 and 2, whose maximum temperatures T_{m1} and T_{m2} ($T_{m1} < T_{m2}$) are close enough to each other as to overlap. After the whole curve has been obtained, a second thermal cycle is started, but we first discharge the lower-temperature peak 1 by heating until T_a ($T_{m1} < T_a < T_{m2}$), then cool the sample again and finally obtain the discharge peak 2, which is nearly pure.

The second cleaning method, specific to TSDC measurements, is due to Bucci et al. (Fig. 1.7).

It consists of first polarizing the material at temperature T_b such that $T_{m1} < T_b < T_{m2}$, so that the dipoles of type 1 are polarized at saturation while the dipoles of type 2 remain practically distributed at random: the resulting TSDC curve will then show only peak 1. Peak 2 can also be isolated in a similar way by using $T_c > T_{m2}$ and removing the field at a temperature T_d such that $T_{m1} < T_d < T_{m2}$.

We have considered in this paragraph only the potential of TSL/TSC techniques in trap level spectroscopy as well as the main problems associated with their application.

References

1. A. Rose, Concepts in Photoconductivity and Allied Problems, Interscience, New York, 1963.
2. Thermally stimulated relaxation in solids, in: P. Braunlich (Ed.), Topics in Applied Physics, vol. 37, Springer, Berlin/Heidelberg, New York, 1971.
3. R. Williams, J. Appl. Phys. 37 (1966) 3411.
4. D.V. Lang, J. Appl. Phys. 45 (1974) 3014.
5. J.C. Carballes, J. Lebally, Solid Stat. Commun. 6 (1968) 167.
6. N.F.J. Mathews, P.J. Warter, Phys. Rev. 144 (1966) 610.
7. I. Broser, R. Broser-Warminsky, Ann. Phys. 6F (1955) 361.
8. M.C. Druver, G.T. Wright, Proc. Phys. Soc. London 81 (1963) 141.
9. F. Urbach, Wien. Berichte (11, a) 139 (1930) 363.
10. J.T. Randall, M.H.L. Wilkins, Proc. R. Soc. (London) A184 (1945) 366.
11. R.H. Bube, Phys. Rev. 83 (1951) 1105.
12. J.G. Simmons, G.W. Taylor, Phys. Rev. B5 (1971) 1619.
13. R.A. Greswell, M.M. Perlman, J. Appl. Phys. 41 (1970) 2365.
14. C. Bucci, R. Fieschi, Phys. Rev. Lett. 12 (1964) 16.
15. P. Braunlich, Thermally Stimulated Relaxation in Solids, Topics in Applied Physics, vol. 37, Springer, Berlin/Heidelberg/New York, 1979.

2 Thermally Stimulated Depolarization Currents in Amorphous Chalcogenides

Contents

2.1 Background

Chalcogenide glasses are oxygen-free inorganic glasses containing one or more kind of chalcogen elements. We follow Mott and Devis (*Electronic Processes in Non-Crystalline Materials*) and Borisova (*Glassy Semiconductors*) for the definitions of "amorphous" and "glass." That is, the amorphous material is a noncrystalline solid, and the glass is amorphous material produced through melt quenching. Chalcogenide glasses based on sulfide, selenide, and telluride alloys in binary and multicomponent systems have evolved much interest in terms of the understanding of basic physics of noncrystalline solids as well as for the development of various semiconductor devices [1–4]. Various unique phenomena are inherent for these glasses, and the most intriguing among them is photodarkening. The preceding phenomena cannot be found in crystalline chalcogenides or in any other amorphous semiconductors and are an interesting subject for fundamental research in the field of disordered materials. Reasonably, they are attractive and promising materials for various applications in modern technologies: optical memory devices, X-ray flat-panel detectors, xerography, and so forth [2–5]. Although a great number of studies have been undertaken to understand the characteristics of electronic and optical properties of noncrystalline semiconductors in the past three decades, many of the properties are still not clear. Some of these properties are fairly well understood, while some are still matter of

Trap Level Spectroscopy in Amorphous Semiconductors. DOI: 10.1016/B978-0-12-384715-7.00002-4

debate. In this context, the study of intrinsic trapping states, especially in the mobility gap of Se-based amorphous semiconductors, as we believe, is extremely important due to their technological applications.

TSC and TSDC are well-known techniques for obtaining data on the trapping levels of crystalline semiconductors, and the techniques has also been applied successfully to amorphous semiconductors [6–18]. The TSDC provides researchers today with an active arena of technological as well as fundamental study. On the fundamental front, TSDC provides a powerful framework for understanding the band gap structure and properties of amorphous materials. The main attraction of TSDC as an experimental method for the study of defects in high-resistance solids was, for many years, its apparent simplicity. Trapping levels in the band gap determine the fundamental electronic properties of both phases. In conventional TSC measurements, difficulties arise if the thermal excitation of equilibrium carriers becomes comparable with the excitation of trapped nonequilibrium carriers. In this situation the TSC signal appears in the best case only as a shoulder on the dark current–temperature curve. One of the main difficulties in observing TSC in amorphous semiconductors is the small magnitude of the TSC currents [11, 14–18]. In most chalcogenide materials—and this may even be considered a universal property—the TSC measurements yield no peak. As for the amorphous ones, it is important to note that no universal method is currently known to detect the entire spectrum of trapping levels in the mobility gap. This is the reason investigators employ several complementary methods. Among these methods are those that are convenient for the study of shallow and deep trapping levels. As for the former, nonisothermal relaxation techniques and TOF measurements seemed to be "suitable"; as for the latter, xerographic spectroscopy is the most frequently employed [5]. Both methods have advantages and disadvantages, and it may often be useful to apply the methods listed earlier to the same specimen. TSDCs allow studying a relaxation processes attributed to relatively shallow trapping levels. The disadvantage in measuring TSDC is the fact that the signals detected are very low and, sometimes, can be observed only in a relatively small temperature interval. At the same time, the advantage is that the TSDC method is inherently more sensitive than other methods, and the resolution is usually much better. In addition, these measurements may be classified as nondestructive.

This section contains a review of results on the extensive study of defect states in the mobility gap of amorphous As- and Sb-containing chalcogenide semiconductors by relaxation technique. For extracting typical features, elemental selenium and simple compositions with relatively low content of arsenic and antimony are exemplified as possible. We will try to attribute TSDC peaks to charge carriers released from the respective trapping levels in the band gap of these materials.

Usually, two basic types of relaxation techniques are used:

a. *Isothermal relaxation*: the perturbation is implemented at a constant temperature.
b. *Nonisothermal relaxation*: the system is perturbed at a sufficiently low temperature to reduce the probability of establishing a new statistical equilibrium. Subsequently, the temperature is increased according to a well-controlled heating program $T(t)$, and the relaxation of the system can be monitored as a function of temperature and time.

In this study, we emphasized the nonisothermal relaxation technique.

The occurrence of TSDC during a thermal scan of a previously excited ("perturbed") material is probably the most direct evidence we have for the existence of electronic trap levels in the band gap of these materials. The main attraction of TSDC and related techniques as experimental methods for the study of the trapping levels in high-resistance semiconductors was their apparent simplicity. A TSDC spectrum (for historical reasons, frequently referred to as a "glow curve") usually consists of a number of more or less resolved peaks in current versus temperature dependence. The latter, in most cases, may be attributed to a species of traps.

Because the escape probability of carriers from trapping states is proportional to $\exp(-E_t/kT)$, the location of glow peak on the temperature scale provides information on the value of the thermal activation energy E_t. Hence, a glow curve represents a spectrum of energy required for carriers to be released from various traps in material.

2.2 Theoretical Background

At the very beginning, as far as amorphous materials are concerned, it must be remembered that the validity of a pure trapping model is still a moot point. At the same time, in such nonperiodic structures (long-range order is absent), the immobilization of charge carriers for long periods necessarily indicates the presence of traps. In addition, it should be noted that although chalcogenide glasses may have traps distributed throughout the mobility gap, it appears justifiable to use the single-trap approach to calculate the trapping parameters (especially activation energy) of the materials under study. Such a model based on simplified version of the relaxation kinetics, as we believe, may be the commonly employed procedure in other complicated cases.

Consider first a single-trap level of density M at energy E_t in the mobility gap ΔE of a p-type semiconductor. The level capture cross section is S and the concentration of trapped carriers m. We assume that initially, the sample is at a uniform temperature T_0, low enough ($T_0 \approx 80\,\mathrm{K}$) to prevent thermal emission of holes from its respective trap. Increasing the temperature according to a heating program $T(t)$ leads to release of trapped carriers until thermodynamic equilibrium is reached again at some higher temperature. Further, we assume the density of equilibrium carriers p_0 substantially less than those released from traps p_t; besides neglecting diffusion and carrier recombination or generation, the excess charges can be considered as uniformly distributed and concentrated in narrow layers close to the electrodes (barrier-type polarization) if at least one of the semiconductor–electrode interfaces can be considered as fully blocking (insulated electrode). The discharge of a photoelectret, i.e., amorphous material previously polarized by the photoelectret effect, may be expressed in the form [8–10]

$$Q(T) = Q_0 \exp\left[\frac{4\pi e\mu}{\chi v_T}\int_{T_0}^{T} p(T)\,\mathrm{d}T\right] \tag{2.1}$$

where Q_0 is the initial charge, μ the microscopic mobility, χ the permittivity of the material, p the concentration of free holes in the valence band, and ν_T the heating rate. The current induced in the measuring circuit by the internal field of photoelectret during heating (irreversible scan $T(t)$) is described by the equation

$$I(T) = \frac{4\pi e\mu}{\chi} Q_0 p(T) \exp\left[-\frac{4\pi e\mu}{\chi\nu_T}\int_{T_0}^{T} p(T)\,\mathrm{d}T\right] \tag{2.2}$$

We may write

$$p(T) = p_0(T) + p_t(t) \quad \text{with } p_0(T) = C\exp\left[-\frac{\Delta E}{2kT}\right] \tag{2.3}$$

where C is the preexponential factor of the dark conductivity and k the Boltzmann constant.

For states distributed in energy or for N discrete states very close to each other, equation (2.3) transforms [8]:

$$p(T) = C\exp\left[\Delta E/kT\right] + \sum_{i=1}^{N} p\left[t_i(T)\right] \tag{2.4}$$

The p_0 and p_t ratio in equation (2.3) determines which of two factors—namely, equilibrium or nonequilibrium (due to emission from traps) carriers—dominate in the relaxation process. That is, the depolarization current contains two maximum: one is related to release of carriers from trap; the origin of the other lies in the change of conductivity with temperature [14–18]. Although only one of the peaks mentioned contains information about trap parameters, it is possible to discriminate between simultaneously occurring processes, e.g., thermally stimulated depolarization and thermally stimulated dielectric relaxation.

For p_t to be defined, it is necessary to solve rate equations [19]

$$\frac{\mathrm{d}(m+p)}{\mathrm{d}t} = \frac{p(t)}{\tau} \tag{2.5}$$

$$\frac{\mathrm{d}m(t)}{\mathrm{d}t} = -\alpha(t)m(t) + \gamma\left[M - m(t)\right]p(t) \tag{2.6}$$

Here $\alpha = \gamma N_v \exp[-E_t/kT]$.

The expressions (2.1)–(2.6) are similar to those describing TSC processes obeying first-order kinetics and represent an asymmetrical glow curve the amplitude of which is a function of heating rate.

Detailed solutions have not been discussed in the literature so far. The time dependence of the rate equations (2.4) and (2.5) is replaced by the temperature dependence via the heating program, which is taken to be linear $T(t) = T_0 + v_T t$, where $v_T = dT/dt$ is the heating rate. Retrapping can be neglected in a sample at high electric fields. From equations (2.5) and (2.6), we have

$$\frac{dp}{dT} = \frac{dm}{dT} + \frac{p}{v_T \tau} \tag{2.7}$$

$$\frac{dm}{dT} = \frac{\alpha(T)\, m(T)}{v_T} \tag{2.8}$$

and for

$$p_t(T) = \tau m(T) \gamma N_v \exp(-E_t/kT) \tag{2.9}$$

where $\gamma = \langle V \rangle S$, V is the average thermal velocity of charge carriers, S the capture cross section (for neutral, attractive, or repulsive centers), and N_v the density of states in the valence band. It is obvious that the only parameters markedly affecting the peak amplitude and position will be the heating rate v_T in combination with the characteristic frequency factor γN_v and the activation energy E_t.

Equation (2.2) is solved numerically with parameters ΔE, C, μ, E_t, and γN_v given. It was found to be a good approximation for a wide range of physically reasonable trapping and other parameters, typical for a wide range of amorphous semiconductors:

$$1.0\,\text{eV} \leq \Delta E \leq 2.0\,\text{eV}; \quad 10^{18} \leq C \leq 10^{22}\,\text{cm}^{-3}; \quad \mu \approx 10^{-3}\text{–}10^{-5}\,\text{cm}^2/\text{Vs}$$

$$0.2\,\text{eV} \leq E_t \leq 0.8\,\text{eV}; \quad 10^{7} \leq \gamma N_v \leq 10^{13}\,\text{s}^{-1}$$

The capture cross section of a trap is largely determined by its charge state. Values reported in the literature (see Refs. [9, 18] and references therein) span the range 10^{-15} to$10^{-12}\,\text{cm}^2$ for Coulomb-attractive centers, 10^{-17} to$10^{-15}\,\text{cm}^2$ for neutral centers and down to $10^{-22}\,\text{cm}^2$ for Coulomb-repulsive centers.

Some examples of equation (2.2) numerical solutions are shown in Fig. 2.1. It should be emphasized here that numerical solutions of equation (2.2) exhibit a greater variety of shapes, peak positions, and magnitudes than shown.

Although two peaks of comparable amplitude are presented (see Fig. 2.1), only the first, denoted as M_1, is actually related to the carriers release from trap, the second, denoted as M_2, is connected with dark conductivity variation with temperature (DC conductivity-determined relaxation peak related to the movement of equilibrium carriers).

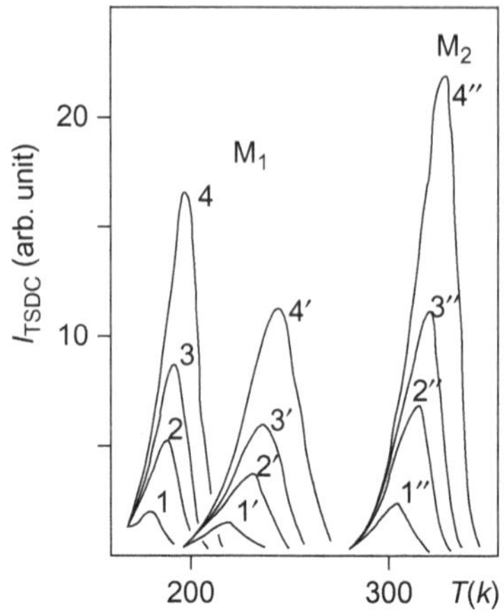

Figure 2.1 Numerical solutions of equation (2.2) for the case $E_t = 0.4\,\text{eV}$ and $\gamma N_v = 10^7$ and $10^9\,\text{s}^{-1}$ (curves 1–4 and 1″–4″, respectively). Heating rate is 0.1, 0.3, 0.5, and 1.0 K/s (curves 1–4, respectively). Also shown are the dielectric relaxation currents (maximum M_2) for the same heating rates (curves 1″–4″, respectively; $\Delta E = 1.8\,\text{eV}$, $C = 10^{22}\,\text{cm}^{-3}$).

Rate equations predict the influence of the heating rate on amplitude as well as on the position of the peak: when the heating rate increases, the initial polarization has to be released in a shorter time. As a result, the peak increases and shifts to a higher temperature. In fact, both peaks, M_1 and M_2, change with heating rate in a similar manner, but only the first has its temperature location (T_m) dependent on the values of E_t and γN_v at the given heating rate.

Such behavior of the TSDC peak is typical. On the other hand, the DC conductivity-determined relaxation peak shifts to lower temperatures when the band gap and/or preexponential factor decreases (Figs. 2.2 and 2.3).

Thus, variation of the conductivity value σ, mobility gap ΔE, and the activation energy of trap let us establish the range where TSDC peaks directly related to traps can be observed "safely." For example, we can detect only traps with $E_t \leq 0.6\,\text{eV}$ if the dielectric relaxation peak locates close to the room temperature.

Illustrative for this statement are TSDC experimental results on $Cu_x(As_2Se_3)_{1-x}$ glasses. Essential increasing of dark conductivity and decreasing of its activation energy E_σ with Cu content is characteristic for these glasses. Accordingly, conductivity-determined depolarization maximum shifts to lower temperatures (Table 2.1). As for the peak associated with traps, those may be observed only in $Cu_{0.05}(As_2Se_3)_{0.95}$. In $Cu_x(As_2Se_3)_{1-x}$ glasses with $x > 0.05$, the TSDC maximum observed is simply the result of two peaks overlapping—those determined by equilibrium (dark) conductivity and, on the other hand, those associated with trapping levels.

The presence of a maximum on the TSDC curve, which is determined by equilibrium conductivity, was also confirmed in the extensive article by Agarwal (in particular, for chalcogenide glasses). Moreover, it was shown that the TSDC curves may contain more than one peak, the position and amplitude of which depends on

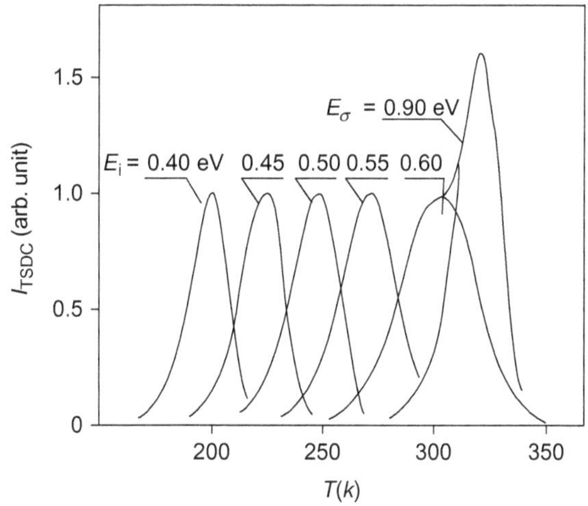

Figure 2.2 TSDC maximum dependence on energy location of trapping level E_i. $\gamma N_v = 10^9\,s^{-1}$, $E_\sigma = 0.9\,eV$, $C = 10^{22}\,cm^{-3}$. The maximum of dielectric relaxation is of greater amplitude.

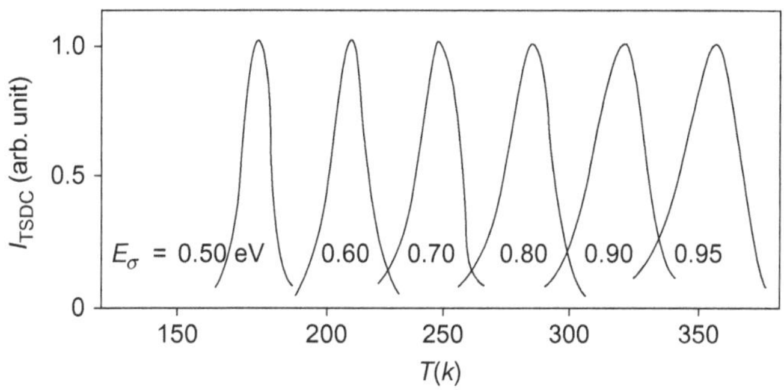

Figure 2.3 Dielectric relaxation maximum as a function of DC conductivity activation energy E_σ.

Table 2.1 Values of E_σ and E_t^* for $Cu_x(As_2Se_3)_{1-x}$ Glasses

Composition	E_σ (eV)	E_t (eV)
As_2Se_3	0.91	0.89
$Cu_{0.05}(As_2Se_3)_{0.95}$	0.60	0.60
$Cu_{0.10}(As_2Se_3)_{0.90}$	0.45	0.40
$Cu_{0.15}(As_2Se_3)_{0.85}$	0.33	0.30

*E_σ and E_t were determined from the temperature dependence of DC conductivity and TSDC curve, respectively.

the size and relative geometrical arrangement of the electrodes with respect to one another [14].

The model presented here has been found to satisfactorily fit experimental data in many cases, particularly for high-resistance, amorphous, chalcogenide semiconductors. We have compared calculated and experimentally observed depolarization currents for $As_{0.5}Se_{0.5}$. In fact, we have observed that two relatively broad maxima of comparable amplitude, curve shape, temperature location, and similar behavior with variation of the heating rate were present on TSDC. The former (peak temperature T_{M1}) is attributed to trapping level, while the latter (peak temperature T_{M2}, such that $T_{M2} > T_{M1}$) is caused by dielectric relaxation current. In fact, one of the major problems of thermally stimulated current measurements is to unequivocally determine the physical origin of the observed current peaks. This is not an easy task, and a great deal of controversy still surrounds the interpretation of TSDC experimental data for most the materials tested. In our case, the origin of the second peak on the depolarization curve is strongly supported by the appearance of single, close-to-room-temperature peak in the TSCD experiment. It involves filling the states located near Fermi level at some temperature (e.g., room temperature) by the application of a strong field and subsequent cooling to a lower temperature with the field applied. Then, the field is removed and the sample heated in the usual manner. The current, measured during heating, is the dielectric relaxation current. It contains information only on electronic states giving rise to dark (equilibrium) conductivity.

2.3 TSDCs in Se-Based Amorphous Semiconductors: Experimental Results

2.3.1 Sample Preparation

The sample preparation is always the same, independent of the concrete measuring technique applied. Usually, the special geometry of the sample may be adapted to the circumstances desired. Special attention was given to homogeneous field distribution in the sample. This allows an unambiguous interpretation of the experimental results. For any leakage currents to be avoided, the sample surfaces were carefully cleaned. Blocking electrodes have been formed simply by pressing small metal lamellae (or foil) to well-polished bulk samples (amorphous film). Highly insulating spacers were put under the lamellae. Thin lead wires were attached to the electrodes and allowed a good electrical contact. The experiments are carried out with different electrode materials and structures with insulated electrodes. In the latter case, insulating layers are inserted between the sample and the measuring electrodes. TSDC measurements were carried out on bulk glass (sample thickness ≈ 0.5 mm) and amorphous film (thickness of ≈ 1–$10\,\mu$m) samples. The heating rate was varied by application of different voltages across the heater. Typical heating rates are in the range 0.1–1.0 K/s. At lower heating rates the current signal become very small, while at higher heating rates the temperature gradient inside the bulk sample causes a signal distortion.

2.3.2 Experimental Arrangement

As a general rule, the detection efficiency expected in conventional TSC experiments will be no more than 0–15% for carrier drift. It is possible, however, to significantly increase the efficiency of TSDC by using an insulating electrode adjacent to one side of the sample (insulating foil). Because the insulating electrode blocks any charge exchange, all image charges previously induced at the noncontacting electrode (due to carriers detrapping from gap states) will be released during the TSDC run; it will also be possible to observe the current resulting from DC conduction. We have initially assumed a well-defined sandwich-cell configuration consisting of a sample that is insulated from the metallic electrodes. The experimental apparatus used for recording TSDC is classic; it essentially composed of a cryostat with a sample holder, a voltage supply, a heater, and a current detector. Standard metal–glass cryostats are used in low-temperature studies of TSDC. The temperature of the films is controlled by mounting a heater inside the sample holder and is measured with a thoroughly calibrated copper–constantan thermocouple. Electrical leads and feedthroughs are designed for minimal leakage currents and stray capacitance. The sample was provided with a digitally controlled voltage supply of extreme stability and low noise level.

TSDC experiments are customarily analyzed assuming the sample behaves "ohmic," i.e., the contacts do not introduce an inhomogeneous distribution of the electric field or carrier density and a uniform bulk density of carriers extend through the entire sample. Experiments were carried out in such a way as to minimize injection effects. Contact configuration was typical for TSDC experiments. Because the currents through the sample are, in almost all cases, extremely small, we have used a sensitive DC ammeter (model U1–15, detection limit $\leq 10^{-15}$ A) with a linear output signal. The simplest way to obtain a record of TSDC is an *X–Y* recorder that displays $I(T)$ and the temperature. The equipment for the extraction of trap-spectroscopic information may be connected with devices for electronic data processing. The experimental errors in E_t determination are less than 2%.

In the TSDC considered here, a sample is cooled to a low temperature ($\approx$100 K) and illuminated with 3×10^3 lx light for a time t_p ($\approx$4 min) in the presence of an applied DC field ($E = 5 \times 10^4$ V/cm). Then, the light and voltage are switched off; the structure is "short-circuited"; and, after a delay period necessary for sample relaxation (to reach equilibrium between the free and the trapped carriers), the sample is heated in the darkness at a constant rate ν_T while the TSDC is measured. We preferred TSDC experiments because of the absence of noise due to a voltage source and the strongly reduced influence of the intrinsic conductivity.

2.4 TSDC in Pure Selenium

Conventional TSC in glassy selenium exhibits monotonic temperature dependence without the distinct "structure" that is characteristic of crystalline analog. The characteristic peaks attributed to traps are usually absent on TSC versus temperature dependencies. Such a behavior is typical for chalcogenide glassy semiconductors.

Although TSC exceeds the dark conductivity, the absence of a well-defined structure of the TSC does not allow identification of the respective trapping levels [17–19]. Therefore, other methods of thermal activation spectroscopy are needed, e.g., TSDCs.

The TSDC on glassy selenium samples start from the "photoelectret" state. The latter is most strongly expressed at blocking electrodes. Thus, and in order to separate the TSDC peak from the DC conduction, we placed a highly insulating layer between the glass specimen and metal electrodes in the measuring cell. Two different materials were used as the dielectric insulator: cleaved mica sheet and Teflon. The two dielectric materials gave essentially the same results.

TSDC measurements on glassy selenium samples display a well-shaped, nearly symmetrical peak at $T_m = 150\,K$. It is a generally known fact that in chalcogenide glasses (amorphous materials), the peaks extend over a wide temperature range (compared to their crystalline analogs). In addition, the peaks are flatter and more symmetrical than expected from the simple expressions (see Section 2.2). For example, $\Delta T'_{max} \approx \Delta T''_{max} = 11\,K$ with $\Delta T'_{max}$ and $\Delta T''_{max}$ being the half-width of the peak from the low- and high-temperature side, respectively. At lower heating rates ν_T, the peak shifts to lower temperatures and is reduced in height, as expected (Fig. 2.4).

The principal trap parameter, the activation energy, can be easily calculated from a single TSDC experiment by means of some characteristic elements of the peak, such as its half-width, inflection point, or initial part of current rise. The most useful one and, in fact, the most frequently exploited, is undoubtedly the initial rise method [20], because it is always easily applied to a previously cleaned peak.

The initial rise method is based on the fact that because the integral term in the $J_D(T)$ function (for details, see Ref. [9]) is negligible at $T \ll T_m$, the first exponential dominates the temperature rise of the initial current, so that

$$I_{TSDC} \sim \text{const}(-E_t/k_T)$$

The activation energy can be determined by plotting $\ln I_{TSDC}$ against $1/T$. In this approximation, a straight line is obtained, the slope of which gives $-E/k$. The

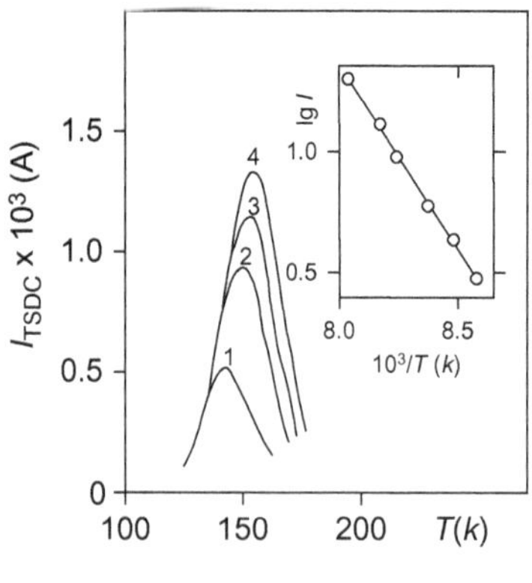

Figure 2.4 Variation in the TSDC for various heating rates ν_T in amorphous selenium: ν_T is 0.08, 0.17, 0.23, and 0.44 K/s (curves 1–4, respectively). The inset shows the Arrhenius plot of TSDC for curve 1.

method has several advantages because it necessitates neither a linear heating rate nor a precise knowledge of the absolute temperature (e.g., it is readily seen that a thermal gradient of 5 K in a sample leads to a relative error $\Delta E/E$ less than 3% at 300 K). It is, however, somewhat limited because use of only the initial part of the curve is permitted, which may force one into the region where the uncertainty in background current is important. On the other hand, the method is still applicable for overlapping relaxation even if it cannot be isolated by cleaning, provided that a series of partial heating, each separated by rapid cooling, is performed up to gradually increasing temperatures that span the whole temperature range of the spectrum.

The activation energy E_t associated with the peak at $T'_m = 150$ K has been calculated by the preceding method, i.e., plotting log I against $1/T$ from the initial rise of the curves. Such a plot for $\nu_T = 0.08$ K/s is shown in Fig. 2.4. The activation energy obtained by this method is 0.22 eV.

An independent check of the correctness of the activation energy E_t can be obtained by measuring the TSDC at a different heating rate ν_T [21]. A number of heating rates are used, and T'_m is determined as a function of ν_T. A plot of $\ln(T'^2_m/\nu T)$ against $1/T'_m$ in such a case yields a straight line [22], and the activation energy is calculated from the slope of that line. Reasonably, the success of this method for the analysis of TSDC curves depends on how much the position of a given peak can be shifted. Therefore, the method will be applicable only when the activation energies are small. It should be noted that the E_t value determined by this method is consistent with that estimated from the initial rise method.

It is necessary to note that the systematic variation in specimen thickness (between 50 and 500 μm) and those of different materials (such as the dielectric insulator) yielded nearly the same peak position and activation energy. This strongly suggests that disturbing contact effects were absent. Moreover, no difference was observed between specimens open-circuited or short-circuited during cooling, nor did the time interval between irradiation and subsequent heating appear to have any effect on the observations. Therefore, it appears that the TSDCs considered here are the true TSDC associated with thermal release of nonequilibrium carriers from localized gap states. The trap of activation energy 0.22–0.24 eV acts as a shallow trap.

In addition to the low-temperature peak at $T'_m \approx 150$ K, a peak at $T'_m \sim 270$ K (not shown in Fig. 2.4) is also present on the TSDC curve. The ratio of amplitudes of these two low- and high-temperature peaks is approximately 5. It is important to note here that the dielectric relaxation current peak (in the depolarization of thermoelectret state experiments) at $T'_m \sim 270$ K is also observed. The activation energy determined for this peak is approximately 0.60 eV. It is essential that this parameter show dependence on electrode configuration and sample thickness. Therefore, a peak with $T'_m \sim 270$ K may be attributed simply to the change in dark conductivity with temperature.

2.5 TSDC in $As(Sb)_xSe_{1-x}$ Alloys

The addition of arsenic to amorphous selenium—even in a relatively low amount (~1 at.%), as for glassy compounds—causes a change in the shape of TSDC curve. As one can see (Fig. 2.5), transition from pure amorphous Se to $As_{0.01}Se_{0.99}$ is accompanied by the broadening of the corresponding TSDC curve and appearance of additional maximum.

The main maximum on the depolarization curve is shifted, respectively, to those observed in a-Se and locates in the range 190–200 K. This maximum is the result of overlapping of two maximums with $T'_m \approx 191$ K and $T'_m \approx 203$ K. In addition, the shoulder at 160 K is also observed. The shoulder and the peak position observed in pure selenium correlate. For the separation of the overlapping maximum, the method of partial thermal cleaning was used. The method is based on release of carriers from traps by heating sequences to temperatures higher than T'_m of the corresponding maximum, rapid cooling, and heating again. The result obtained after applying such a procedure to $As_{0.01}Se_{0.99}$ is shown in Fig. 2.6.

Values of activation energy calculated by the initial rise method are in the range 0.25–0.45 eV. Continuous distribution of localized states in this range was confirmed by an increase of activation energy in a sequence—curves 2–4 (Fig. 2.6).

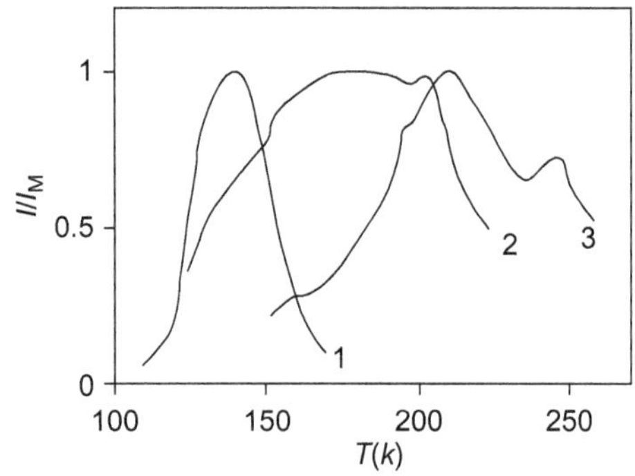

Figure 2.5 TSDCs in glassy Se (1), $As_{0.01}Se_{0.99}$ (2), and $As_{0.15}Se_{0.85}$ (3).

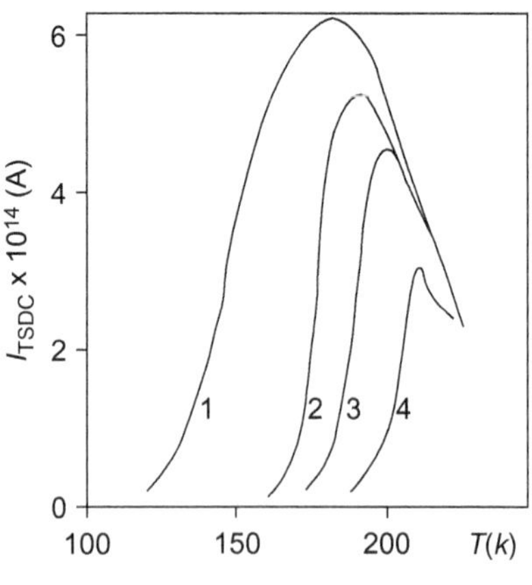

Figure 2.6 Application of thermal cleaning method of TSDC peaks to $As_{0.01}Se_{0.99}$. Curve 1—initial TSDC curve. The temperature in the successive cycles of heating is 166 K (2), 180 K (3), and 199 K (4).

Further increasing the addition of arsenic (up to 6 at.%) does not change the shape of TSDC curve: the latter remains similar to that observed in $As_{0.01}Se_{0.99}$.

In Fig. 2.7, the TSDC observed in As_xSe_{1-x} for As content 8–15 at.% was shown. A peak at $T'_m \approx 210$ K dominates in the depolarization curve. In addition, a peak at $T'_m \approx 244$ K and a shoulder at $T_{max} \approx 197$ K are discernable. Release of carriers from shallow states and subsequent trapping on relatively deeper ones transform the TSDC into curves 3 and 4 (Fig. 2.7). Activation energy determined for peaks with $T'_m \approx 210$ K and $T'_m \approx 244$ K is $E_{t2} = 0.35$ eV and $E_{t3} = 0.45$ eV, respectively. These states are energetically distributed: $\Delta E_{t2} = 0.05$ eV and $\Delta E_{t3} = 0.1$ eV, respectively.

For TSDC in Sb_xSe_{1-x} alloys, we have three peaks M_1, M_2, and M_3, whose temperatures T_{M1}, T_{M2}, and T_{M3} (145, 175, and 195 K, respectively) are close enough to each other that they overlap (Fig. 2.8). The TSDC method allows the easy and quick resolution of the overall spectrum, in our case peaks 1, 2, and 3. There are two ways in which this can be done.

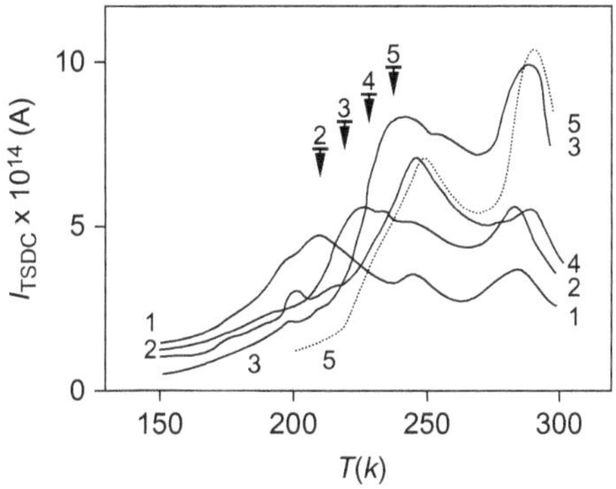

Figure 2.7 TSDC curves of $As_{0.15}Se_{0.85}$ samples (1). Curves 2–5—polarized samples were heated under applied DC voltage to the temperatures shown by arrows.

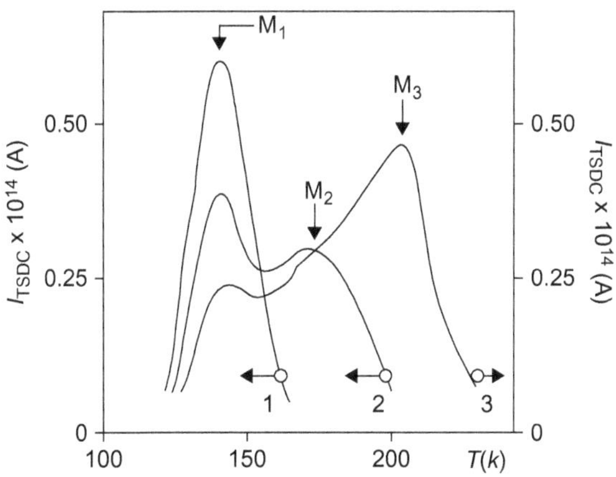

Figure 2.8 TSDC spectra of Sb_xSe_{1-x}, $x = 0$ (1), 0.01 (2), and 0.05 (3).

The first method, common to all nonisothermal techniques, was developed by Greswell and Perlman [22]. Suppose, for example, we have two peaks, 1 and 2, whose temperatures T_{m1} and T_{m2} ($T_{m1} < T_{m2}$) are close enough to each other as to overlap. After the whole TSDC curve has been obtained, a second thermal cycle is started. However, instead of recording the discharge at one time, we first discharge the lower-temperature peak 1 by heating until T_a ($T_{m1} < T_a < T_{m2}$), then cool the sample again and finally obtain the discharge of peak 2, which is "pure" or nearly "pure."

The second cleaning method, specific to TSDC measurements and which may be applied in our case, is due to Bucci et al. [23]. It consists of first polarizing the material at a temperature T_b such that $T_{m1} < T_b < T_{m2}$ for an interval of time $\tau_1(T_b) \ll \tau_2(T_b)$; the resulting TSDC curve will then show only peak 1. Peak 2 can also be isolated in a similar way by using $T_c > T_{m2}$ and removing the field at temperature T_d such that $T_{m1} < T_d < T_{m2}$.

Such an effective way of separating the aforementioned peaks is illustrated in Fig. 2.9.

The activation energies determined by these methods were $E_{t1} = 0.22\,\text{eV}$, $E_{t2} = 0.34\,\text{eV}$, and $E_{t3} = 0.45\,\text{eV}$. Note that some of the traps in the materials studied were "visible" in TOF and xerographic experiments [5, 17, 19].

The results taken as a whole reveal the existence of at least three different trap species in the band gap of $Sb(As)_xSe_{1-x}$ noncrystalline semiconductors. These species are located at energies 0.22, 0.34, and 0.45 eV, respectively, below the conduction band edge and control the electron transport properties of the material. It seemed that Sb and As introduce a new set of detectable charge-carrier traps.

Space considerations preclude a detailed discussion of similar trapping levels in other high-resistance amorphous chalcogenides.

Finally, we are convinced that the future development of TSC and TSDC methods as investigative tools of electronic properties of amorphous semiconductors will

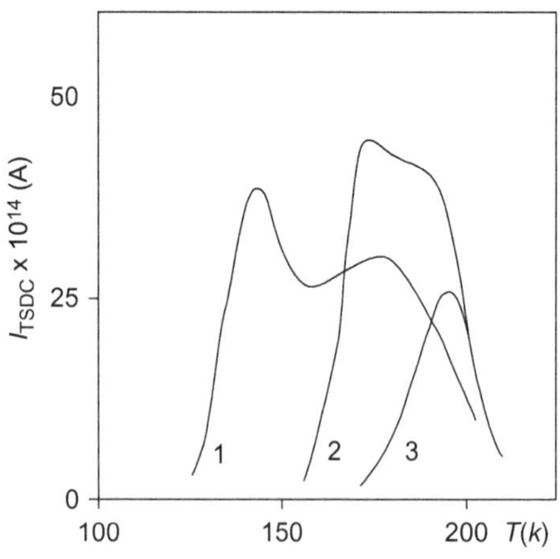

Figure 2.9 Thermal cleaning procedure (of TSDC) applied to $Sb_{0.01}Se_{0.99}$. Curve 1—the whole TSDC curve. For curves 2 and 3, $T_a = 165\,\text{K}$ and $T_b = 195\,\text{K}$ (see the text).

benefit from similar skillful studies via TOF and xerographic measurements [24–26]. It seems to be important that in extensive work [26], electron TOF transient photocurrents have been investigated in stabilized a-Se as a function of electric field, annealing, aging, and alloying with As and doping with Cl. Moreover, the distribution of localized states in stabilized a-Se has been investigated by comparing the measured and calculated transient photocurrents. It should be noted also that the theoretical analysis of multiple-trapping transport has been done for continuous density of states. Undoubtedly, these results are in full agreement with our interpretation, although some deviation in localized states density is observed, possibly because of more precise TOF technique applied.

References

1. M. Popescu, J. Non-Cryst. Solids 352 (2006) 887.
2. V. Kolobov (Ed.), Photoinduced Metastability in Amorphous Semiconductors, Wiley-VCH, Weinheim, 2003.
3. J. Singh, K. Shimakawa, Advances in Amorphous Semiconductors, Taylor and Francis, London, 2003.
4. G. Lucovsky, M. Popescu, Non-Crystalline Materials for Optoelectronics, INOE Publishers, Bucharest, 2004.
5. S.O. Kasap, Boom, in: A.S. Diamond, D.S. Weiss (Eds.), Handbook of Imaging Materials, second ed., Marcel Dekker, Inc., New York, 2002.
6. D. Kumar, S. Kumar, Chalcogenide Lett. 1 (2004) 49.
7. E. Skordeva, J. Optoelectron. Adv. Mater. 3 (2001) 437.
8. E.I. Adirovich, Sov. Solid State Phys. 3 (1961) 2048 (in Russian).
9. P. Braunlich, P. Kelly, J.P. Fillard, Thermally Stimulated Relaxation in Solids (Top. Appl. Phys. 37), Ed. P. Braunlich, Springer, Berlin, 1979.
10. Yu. Gorokhovatsky, G. Bordovskij, Thermally Activational Current Spectroscopy of High-Resistance Semiconductors and Dielectrics, Nauka, Moscow, 1991 (in Russian).
11. R.A. Street, A.D. Yoffe, Thin Solid Films 11 (1972) 161.
12. B.T. Kolomiets, V.M. Lyubin, V.L. Averjanov, Mater. Res. Bull. 5 (1970) 655.
13. T. Botila, H.K. Henish, Phys. Status Solidi (a) 36 (1976) 331.
14. S.C. Agarwal, Phys. Rev. B 10 (1974) 4340.
15. S.C. Agarwal, H. Fritzsche, Phys. Rev. B 10 (1974) 4351.
16. P. Muller, Phys. Status Solidi (a) 67 (1981) 11.
17. V.I. Mikla, Dr. Sci. Thesis, Institute of Physics, National Academy of Sciences, Kiev, 1998.
18. A.A. Kikineshi, V.I. Mikla, I.P. Mikhalko, Sov. Phys. Semicond. 11 (1977) 1010.
19. V.I. Mikla, I.P. Mikhalko, Yu. Nagy, J. Phys. Condens. Matter 6 (1994) 8269.
20. G.F. Garlick, A.F. Gibson, Proc. Phys. Soc. 60 (1948) 574.
21. A.H. Bohun, Can. J. Chem. 32 (1954) 214.
22. R.A. Greswell, M.M. Perlman, J. Appl. Phys. 41 (1970) 2365.
23. C. Bucci, R. Fieschi, G. Guidy, Phys. Rev. 148 (1966) 816.
24. S.O. Kasap, C. Juhasz, J. Non-Cryst. Solids 72 (1985) 23.
25. C. Juhasz, S.O. Kasap, J. Phys. D Appl. Phys. 18 (1985) 721.
26. K. Koughia, Z. Shakoor, S.O. Kasap, J.M. Marshall, J. Appl. Phys. 97 (2005) 033706.

3 Carrier Transport Processes in Amorphous Solids

Contents

3.1 Background

During the past decades, increasing use was made of amorphous semiconductor films, which not only offer the advantage of reduced cost, but are readily produced as large-area elements of the type required in applications.

The successful use of noncrystalline semiconductors in various applications depends on the development of our understanding of electronic conduction in these materials to a standard comparable to that which presently exists in the case of their crystalline counterparts.

For many years, during and after the development of the modern band theory of electronic conduction in crystalline solids, it was not considered that amorphous materials could behave as semiconductors. The occurrence of bands of allowed electronic energy states, separated by forbidden ranges of energy, to become firmly identified with the interaction of an electronic waveform with a periodic lattice. Thus, it proved difficult for physicists to contemplate the existence of similar features in materials lacking such long-range order.

In hindsight, all of the necessary clues were available, including for example the ability of conventional glasses to transmit light of sufficiently long wavelength, and the photoconductive behavior of solids such as amorphous selenium. The blind spot that had developed was not exposed until the mid-1950s, when Kolomiets and coworkers ushered in the current area of interest with their reports of semiconducting behavior in various chalcogenide glasses [1].

Trap Level Spectroscopy in Amorphous Semiconductors. DOI: 10.1016/B978-0-12-384715-7.00003-6

Stimulated by a variety of commercial applications in fields such as xerography, solar energy conversion, thin-film active devices, and so forth, international interest in this subject area has increased dramatically since these early reports. The absence of long-range order invalidates the use of simplifying concepts such as the Bloch theorem, the counterpart of which has proved elusive for disordered systems. After more than a decade of concentrated research, there remains no example of an amorphous solid for the energy band structure, and the mode of electronic transport is still a subject for continued controversy.

It is necessary to note that in addition to the problems associated with the theoretical analysis of electronic conduction in disordered systems, experimental techniques of established success in the study of crystalline semiconductors have proven to yield ambiguous results when applied to amorphous solids.

It is possible to envisage a number of different processes by which a charge carrier might move between locations within amorphous semiconductors. For simplicity, we consider the case of electron carriers, but the concepts developed are equally applicable to hole transport [2].

It is possible for electrons of energy E_A (which is ~0.1 eV above the band edge) to move within extended electronic states. Applying the consideration of conventional band theory, one can estimate the minimum value of mobility μ_0 for electrons in such states as

$$\mu_0 = 100\ \text{cm}^2/\text{V/s} \tag{3.1}$$

For mobility smaller than this, the calculated free path between scatterings becomes smaller than

$$\mu_0 = 10\ \text{cm}^2/\text{V/s}$$

and the mean free path becomes less than interatomic spacing. Cohen [3] has computed a lower limit for extended state transport of

$$\mu_0 = 1\ \text{cm}^2/\text{V/s} \tag{3.2}$$

Electronic conduction in crystalline semiconductors (except for the case of extremely high doping levels or very low temperatures) invariably involves motion in extended states. However, because of the high densities of defect centers, the possibility exists for transport by direct tunneling between localized states.

Even in the presence of very high densities of localized states, μ_h is not expected to exceed a value of $10^{-2}\,\text{cm}^2/\text{V/s}$.

Considering the case of electronic transport at a specific energy, the carrier mobility is envisaged as decreasing rather sharply in the vicinity of the boundary between extended and localized states. Consequently, this dividing energy has been termed a "mobility edge."

In the case of material with a significant concentration of localized states, it is possible to assume that transport of a carrier over any macroscopic distance will involve motion in states confined to a single energy. Here it is necessary to note that a particularly important departure from this limiting situation is (according to Rose [4]) a trap-limited band motion. In this case, transport of carrier via extended states is repeatedly interrupted by trapping in localized states. The macroscopic drift mobility μ_d for such a carrier is reduced from the value for free carriers, by taking into account the proportion of time spent in traps. Under steady-state conditions, we may write

$$\mu_d = \mu_0[\tau_t(\tau_t + \tau_r)^{-1}] = \mu_0[n_t(n_t + n_r)^{-1}] \tag{3.3}$$

where τ_t and τ_r are, respectively, the mean free carrier trapping time and the mean trapped carrier release time, and n_t and n_r are, respectively, the steady-state densities of free and trapped carriers.

For the simplest case of a single set of localized states situated at a particular energy E_1, the trap-limited drift mobility of carriers moving in extended states at E_A is readily computed from equation (3.3). If the effective density of extended states at E_A is N_A and the trap concentration is N_t, then we may write

$$n_t \approx N_A \exp[-(E_A - E_t)/kT] \quad n_1 \approx N_t \exp[-(E_1 - E_t)/kT] \tag{3.4}$$

3.2 Experimental Techniques for the Measurement of Carrier Mobility

In crystalline semiconductors, the most common technique for the measurement of carrier mobility involves the Hall effect. However, in noncrystalline materials, experimental data are both fragmentary and anomalous (see, for example, Ref. [5]). Measured Hall mobility is typically of the order of $10^{-1} - 10^{-2}$cm^2/V/s and is frequently found to exhibit an anomalous sign reversal with respect to other properties providing information concerning the dominant charge carrier. Thus, apart from some theoretical interest, the Hall effect measurements are of minimal value in the study of macroscopic transport in amorphous semiconductors.

A second common technique in the case of crystalline semiconductors is the procedure developed by Haynes and Shockley [6] for the measurement of carrier drift mobility (Fig. 3.1(a)). In its original form, as applied to the examination of specimens of relatively high electrical conductivity, the technique is restricted to the study of minority carrier transport. Excess packets of majority carriers rapidly spread out during transit so that information concerning their transit time along the specimen is lost. Because this spreading occurs in a time of the order of dielectric relaxation time, it will take place more slowly in materials of low electrical conductivity, as is the case for a range of amorphous semiconductors of interest. Against this, amorphous semiconductors are characterized by carrier drift mobility, which is typically

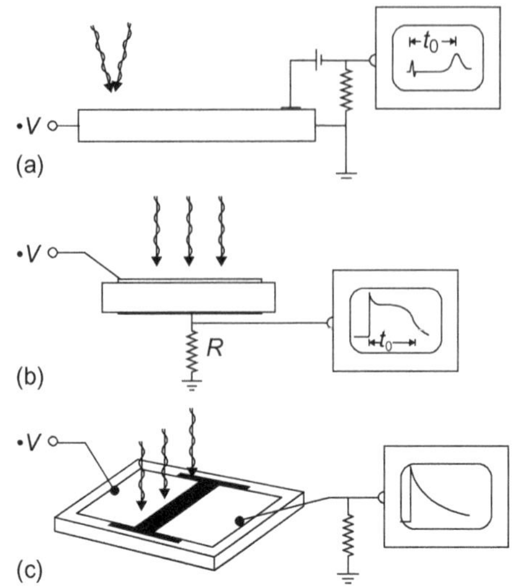

Figure 3.1 Various experimental techniques for the examination of carrier drift mobility. (a) The Haynes–Shockley technique, (b) the Spear technique, and (c) the transient photodecay technique.

much lower than in crystalline semiconductors, so that transit times must be reduced to acceptable short values if the technique is to be of value.

The experimental configuration of Fig. 3.1(b), developed by Spear [7], achieves the preceding objective and has been widely used in the study of carrier transport in low-mobility materials. Excess charge carriers of both signs are created in equal concentrations, close to the upper surface of a thin specimen film. A short-duration excitation pulse of strongly absorbed photons or electrons is normally used for this purpose. Depending on the polarity of the electric field applied across the specimen, carriers of one sign drift across the film, inducing a signal in the series resistor R. This signal will fall to zero when all carriers have completed their transit, allowing an identification of the mean carrier transit time, and thus of the drift mobility. Note that the carrier of either sign may, in principle, be examined, because the dielectric relaxation time in suitable materials is much longer than the transit time through the thin film.

Under the idealized conditions of an infinitely thin sheet of excess carriers drifting through the specimen with zero dispersion, and with suitable blocking electrodes to prevent undesirable re-injection of charge, the voltage induced across the sampling resistor R would fall abruptly to zero at a time equal to the transit time of the carriers. In practice, of course, a spread of arrival times always occurs due to diffusion and other effects so that a reasonably well-formed signal might be of the type shown in Fig. 3.1(b). With such a pulse shape, it remains comparatively easy to identify a mean transit time for the excess carriers, from which the average drift mobility may be computed. However, as noted earlier, the "transit pulses" observed in the study of amorphous semiconductor films often exhibit such an anomalous degree of dispersion that an identification of the drift mobility becomes both difficult and of uncertain significance.

The reason for the interest that has been directed toward measurements of drift mobility in amorphous semiconductors is twofold. First, the performance of many amorphous semiconductor devices such as thin-film transistors and solar cells depends significantly on the achievement of acceptable values of carrier mobility. Second, it may be possible to utilize such measurements in the determination of carrier transport mechanism and localized state distributions in the materials concerned. For example, consider the case of trap-limited band transport in the presence of a single set of localized centers. Given experimental data over a sufficiently wide range of temperature, it is possible to determine the free carrier mobility μ_0, the energy depth of the localized states (from the activation energy of μ_d in the low-temperature regime), and the density of the trapping centers (given a reasonable estimate for the effective density of bad states N_c). However, the validity of such conclusions naturally depends on the successful identification of the dominant trapping mechanism. The correctness or otherwise of assumptions concerning the conduction mechanism occurring in materials under examination has been a subject of much controversy in the study of amorphous semiconductors.

In the case of those noncrystalline solids that are of sufficiently high electrical conductivity that dielectric relaxation proscribes the application of the transit time outlined earlier, the experimental configuration displayed in Fig. 3.1(c) may be of the value. Here, carriers of both species are excited in equal and uniform concentration across the active area of a specimen film fitted with coplanar electrodes. For step-function illumination, the rate of increase of photocurrent with time is linearly proportional to the carrier generation rate and the carrier drift velocity (and at times sufficiently short that recombination may be neglected). Thus, under the assumption that one species of carrier dominates the behavior, its mobility may be determined.

Because both types of carrier are created uniformly in equal concentrations, and because "sweep-out" effects associated with the arrival of carriers at the specimen electrodes may be neglected in materials of low mobility, local charge neutrality is maintained and dielectric relaxation effects are not relevant. However, a disadvantage of the technique is that it is impossible to determine the polarity of the dominant carrier unless information from other measurements, such as thermoelectric power, is employed.

A further application of the coplanar cell configuration showed in Fig. 3.1(c) concerns the study of the time dependence of the photocurrent following carrier excitation by means of a short pulse of illumination. This "transient photodecay" technique enables the examination of the interaction of initially free carriers with various localized states. In principle, the decay of photocurrent measured in this manner should (in the absence of recombination effects and phenomena associated with drift close to the surface of a thin film) correspond to the behavior in the initial pre-transit regime of a "TOF" pulse. Because it allows measurements to be performed on very thin films under conditions appropriate to their use in many device applications, and because the photocurrent may be examined over several decades of time without the complications associated with carrier extraction, the technique has become rather popular over recent years.

3.3 Significance of Carrier Transport Data in Various Applications

It is appropriate to review the various major commercial applications of amorphous semiconductor devices and to indicate the extent to which the usefulness of amorphous semiconductors depends on an optimization of their electronic transport properties.

The most familiar application of amorphous semiconductors will, for many readers, be in the field of replication of printed matter. The xerography process, upon which many modern photocopiers are based, involves the ability of an electrostatically charged plate of amorphous chalcogenide (or similar material) to discharge under illumination. Residual charging of illuminated areas is employed in the transfer of "ink" onto the duplicator paper. Naturally, the mobility of photoinduced carriers in the amorphous semiconductor photoreceptor is of central importance in the validity of the process, and considerable commercial effort has been (and is being) devoted to the study of transport in disordered materials suitable for the process.

A second major application of amorphous semiconductors, of comparatively recent emergence, concerns the conversion of solar energy into electrical power. Following the oil crisis of the mid-1970s, appreciable attention has been paid on an international scale to the development of alternative sources of power. For many years, amorphous semiconductors were not considered suitable as photovoltaic cell materials due to the difficulty experienced in doping such materials to produce the necessary p–n junctions. However, in 1975, the first reports of successful doping of amorphous silicon films were published; since then, the technology of a-Si solar cells has shown progressive improvement.

Because fabrication costs show the potential for substantial reduction, and because amorphous silicon films are much more readily created in the form of large-area panels than are their crystalline counterparts, it is anticipated that amorphous devices may command a substantial share of the commercial market, once the necessary development stages have been accomplished. Already, devices such as calculators, television sets, and tape recorders have been produced with amorphous silicon solar panels as the power source, and an experimental solar-powered housing unit has been constructed.

Solar conversion efficiencies for amorphous cells are increasing steadily, but again the performance of devices critically depends on the mobility and lifetime of excess photocarriers in the films.

The fabrication of active semiconductor devices from amorphous semiconductor films is a further application that offers considerable advantages. Thin-film transistors, based on amorphous films of hydrogenated silicon, are under intensive development. Other devices with monostable and bistable switching characteristics have also received considerable interest. Naturally enough, the performance of such devices is intimately related to the transport properties of charge carriers in the materials employed.

From a commercial standpoint, except in the case of mature technology associated with the xerographic process, the present state of development of amorphous semiconductor devices bears a similarity to that existed in the 1950s for their crystalline counterparts. Application has been envisaged and their feasibility demonstrated, but ultimate success depends on an improvement in electronic characteristics to such a level that devices may compete favorable against existing and future alternatives.

3.4 Conventional Dispersion Behavior

As a consequence of random variations in the propagation characteristics of individual charge carriers, an initially discrete packet of carriers will necessary broaden out in profile as it drifts across a specimen. For the carriers that move exclusively in extended states, this dispersion results from statistical variations in scattering processes and may be described in terms of a diffusion coefficient D, which is related to the carrier mobility via Einstein relation

$$D = (kT/e)\mu \tag{3.5}$$

The broadened packet of carriers assumes a Gaussian profile, with deviation from the mean position of

$$\Delta l = (2Dt)^{1/2} \tag{3.6}$$

The mean elapsed time for a carrier to complete a transit across a specimen of width l under the influence of an applied electric field E is

$$t_0 = l/\mu E \tag{3.7}$$

and the spread of transit times of individual carriers will be of the order of

$$\Delta t = \Delta l/\mu E \tag{3.8}$$

Hence, the relative degree of dispersion of the drifting charge packet is of the order of

$$\Delta t/t_0 = \Delta l/l = (2D/\mu)^{1/2}(El)^{-1/2} \tag{3.9}$$

Thus, if measurements are performed on specimens of differing thicknesses, at identical applied fields, then variations in the relative degree of dispersion should be observed, with thicker specimens being characterized by less dispersive transit pulses (Fig. 3.2). It is interesting to note that a number of analysis (see Ref. [8]) have assumed that an increase in electric field in a given specimen will result in an increase in relative dispersion (due to a reduction of the mean transit time). However,

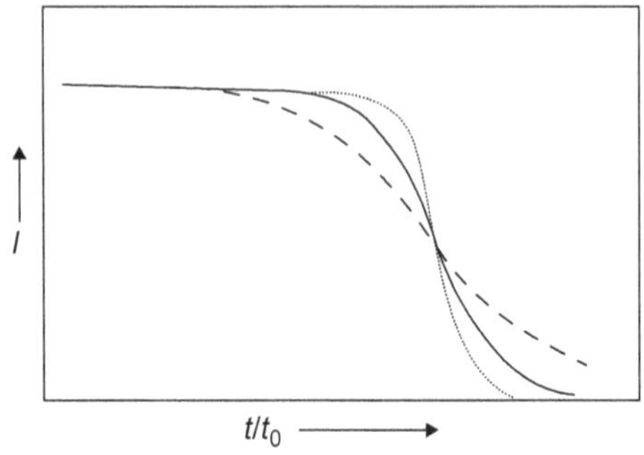

Figure 3.2 Variations in the relative degree of transit pulse dispersion in the case of conventional carrier transport. The specimen thickness is least for the broken curve and greatest for the dotted curve.

equation (3.9) demonstrates that such an increase in field should reduce the degree of relative dispersion.

In circumstances where carrier transport involves isoenergetic hopping between localized states in a regular spatial array, Gaussian dispersive characteristics would again be expected as a consequence of random statistical variations in the dwell times of carriers upon individual sites. Similarly, such dispersion might arise where band transport is interrupted by trapping in a single isoenergetic set of localized states, due to fluctuations in trapped carrier release times.

Under realistic experimental conditions, additional dispersion may well be introduced as a consequence of factors, such as a finite initial width of excess carrier packet or fluctuations in electric field across a specimen film. Most, if not all, of these factors would be expected to yield a relative dispersion that decreases with increasing specimen thickness. In a well-conducted experimental measurement, a relative dispersion of the order of 20% might typically be achieved, giving a transit pulse similar to that shown in Fig. 3.2 (full line) (from which the mean carrier transit time is readily determined).

3.5 Anomalous Dispersive Characteristics

While transit pulses of the preceding form are observed in many crystalline and in some amorphous specimens, the behavior exhibited in other cases differs to a surprising degree.

The study of transport in disordered semiconductors has lead to widespread observation of carrier transit pulses exhibiting an anomalous degree of dispersion. Conventional theory predicts a progressive Gaussian broadening of the drifting charge sheet, typically producing transit pulses of the shape shown in Fig. 3.3.

Figure 3.4(a) displays a pulse obtained during a study of the transport of hole carriers in a-As_2Se_3. It will be observed that the current in the initial regime (prior to

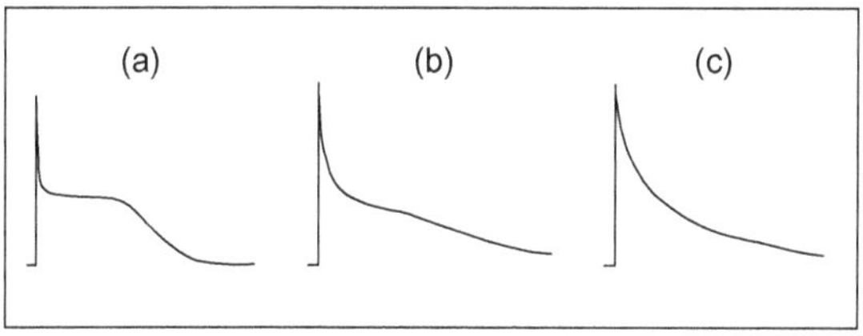

Figure 3.3 Carrier transit pulses as obtained for (a) conventional transport and (b, c) anomalously dispersive transport.

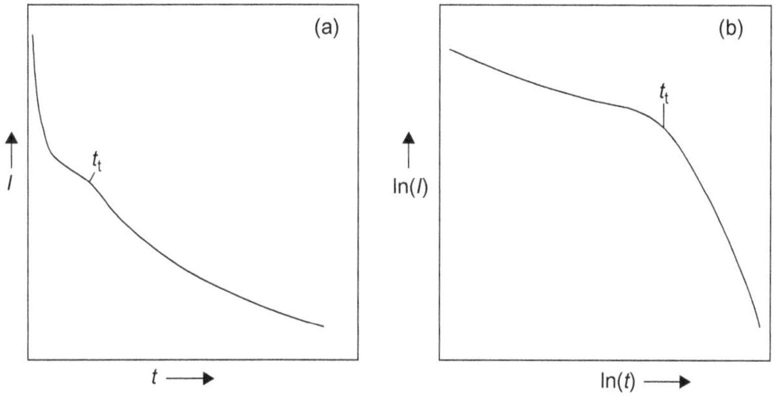

Figure 3.4 Typical transient response characteristic for hole carrier transport in a-As_2Se_3 at room temperature. The behavior is displayed with both linear (a) and logarithmic (b) axes of current and time.

the change of gradient in the vicinity of time t_t, associated with the arrival of the leading edge of the excess carrier packet at the extinction electrode) does not assume a constant value, as is the case in Fig. 3.2. Rather, the average current per drifting carrier would appear to decrease continuously with increasing time over the whole time range of the measurement. Additionally, the spread of arrival times of individual carriers at the extinction electrode, reflected in the behavior at times greater than t_t, is much greater than that expected from conventional theory. Even at times greater than t_t, the magnitude of the current is such as to suggest that a significant number of carriers remain within the specimen. Note that while the transition time t_t introduced earlier has an association with the transit time of the fastest carriers across a specimen, it is distinctly different from the average carrier transit time.

The observation of pulses of the form shown in Fig. 3.4, in the case of hole transport in a-As_2Se_3, led Scharfe [9] to describe the phenomenon as involving carriers possessing a range of drift mobility. However, at that time, the concept was not in itself of great value, because it was difficult to imagine how such a spread of drift velocities could arise in an apparently homogeneous material. Thus, although transit pulses of this anomalous form were observed in various noncrystalline materials during the early 1970s, their origin remained a mystery. Even so, a detailed phenomenological characterization of the phenomenon was achieved during this period, primarily as a consequence of various studies by investigators at the Xerox Research Laboratories.

The phenomenon, which has come to be termed "anomalously dispersive transport," was characterized during this period with respect to the following properties:

(i) *Universality.* Transit pulses obtained from measurements upon a given specimen at various values of applied electric field were suggested to exhibit almost identical degrees of relative dispersion. This universality of transit pulse shape, which clearly contradicts equation (3.5), was also suggested to extend to variations in specimen thickness and (in some cases such as that of hole transport in a-As_2Se_3) to variations in temperature.

(ii) *Algebraic time dependence of excess current.* If anomalously dispersive transit pulses were replotted using logarithmic axes of both current and time, the behavior was found to involve two approximately linear regimes, obtaining for times less than t_t and for times greater than t_t:

$$I_1 \sim t^{-(1-\alpha 1)} \quad (0 < t < t_t) \tag{3.10a}$$

$$I_2 \sim t^{-(1+\alpha 2)} \quad (t > t_t) \tag{3.10b}$$

These algebraic or power-law characteristics were further suggested to interrelate, in sense that the parameters α_1 and α_2 were identical:

$$\alpha_1 = \alpha_2 = \alpha \quad (0 < \alpha < 1) \tag{3.10c}$$

The preceding characterization of anomalously dispersive transit pulses aroused considerable interest, both from a theoretical and an experimental viewpoint. Attention focused on the latter was stimulated by the possibility of using the log–log display technique to identify t_t in cases where the dispersion was such as to obscure any change of gradient in conventionally displayed transit pulses. However, it became necessary to question the validity of such measurements of t_t under conditions where individual carrier transit times vary over such a wide range.

Figure 3.5 displays the behavior for a concentrated system, while Fig. 3.6 shows the corresponding characteristics for a system of sufficient dilution to display anomalous dispersion. In the former case, the carrier packet spreads out with a conventional Gaussian profile, and the average carrier displacement increases linearly with time. Under the latter conditions, the broadening is much more pronounced, and a significant number of carriers remain localized close to the top electrode—even when other members of the distribution have approached or reached the extraction electrode. The immobilized carriers are envisaged as having been trapped in centers that (due to randomness of the system) are more isolated from their neighbors than is the average site. Scher [8] suggests that with increasing time, drifting carriers will encounter progressively more isolated sites with longer and longer release time constants, giving a transient current that decays continuously with time, rather than rapidly approaching some steady-state value.

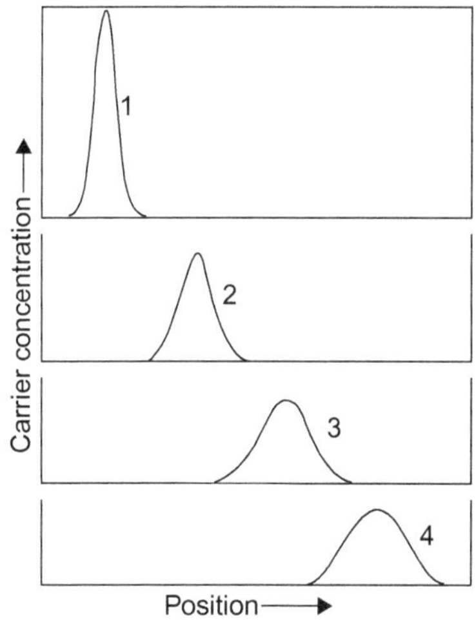

Figure 3.5 Time evolution of the distribution of drifting charge carriers for the case of the conventional dispersion [8].

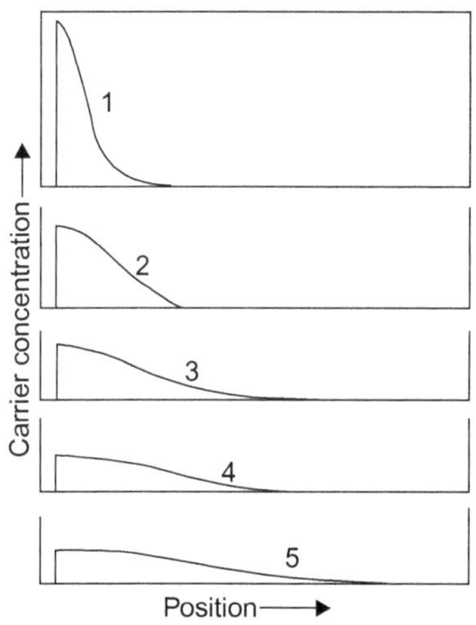

Figure 3.6 Time evolution of the distribution of drifting charge carriers for the case of anomalous dispersion [8].

The first attempts to explain why transit pulses in amorphous semiconductors should exhibit anomalous dispersion appear to have been made in 1971. Marshall and Owen [10] examined the possibility that in such thin specimen films, carriers moving in extended states might experience just a few trapping events during transit. Thus, while some carriers might be trapped several times during trapping, others would cross the specimen without localization, giving a broad spread of individual carrier transit times. Although such a mechanism would yield a considerable degree of dispersion, it was difficult to reconcile it with the apparent characteristics. It was demonstrated by Marshall and Owen [10] that, for an isoenergetic set of traps, an average 10 or more trapping events per transit (n_0) was required if carriers were to approach steady-state average drift mobility of the corresponding equation. Taking n_0 as 3 would yield a high degree of dispersion, but variation about this figure produced by changes in specimen thickness or electric field would lead to marked changes in pulse shape.

The study of the dispersion of photoinjected charge-carrier packets in conventional TOF measurements can provide important information about the electronic and ionic charge transport mechanism in disordered semiconductors [5]. In several materials—among which polysilicon, a-Si:H, and amorphous Se films are typical examples—it has been observed that following photoexcitation, the TOF photocurrent reaches the plateau region, within which the photocurrent is constant, and then exhibits considerable spread around the transit time. Because the photocurrent remains constant at times shorter than the transit time and, further, because the drift mobility μ_d determined from t_t does not depend on the applied electric field, the sample thickness carrier thermalization effects cannot be responsible for the transit time dispersion observed in these experiments.

There are several mechanisms that can lead to the carrier packet and transit time dispersion under quasi-equilibrium transport conditions [11]. First, the propagating carrier packet experiences dispersion arising from the mutual Coulombic repulsion of the carriers. This Coulombic spatial dispersion increases linearly with time and with the amount of injected charge Q_0 as

$$\sigma_C(t) = (\mu Q_0/\varepsilon_0 \varepsilon S)t \tag{3.11}$$

where S is the sample surface area that is illuminated, ε_0 the permittivity of vacuum, and ε the dielectric constant of the material. The corresponding Coulombic dispersion Δt_C of the equilibrium transit time is

$$\Delta t_C = [\sigma_C(t_t)/\mu E] = Q_0 L/(\varepsilon_0 \varepsilon S \mu E^2) \tag{3.12}$$

The Coulombic dispersion has been neglected in almost all TOF analysis inasmuch as it is generally assumed that the internal field is constant under the small signal regime, i.e., $E = V_0/L$, where V_0 is the bias voltage. Under typical experimental small signal conditions, the fractional injected charge can be about 1–10% of the electrode charge, which means that there will be an inevitable contribution to the observed dispersion from the mutual Coulomb repulsion of injected carriers. However, this effect can be eliminated by examining experimentally obtained dispersion as a function of injected charge Q_0 and linearly extrapolating it to $Q_0 = 0$.

Another possible origin of the transit time dispersion is the conventional thermal diffusion of charge carriers. Dispersion (in this context) is discussed in terms of square root of the variance of the charge-carrier concentration profile at a given time; the carrier concentration profile propagates along the field direction with a mean drift velocity that is μE. In this case, the carrier packet shape is described by a Gaussian function with the root mean square spread, or dispersion, σ_d, being proportional to $t^{1/2}$

$$\sigma_d(t) = (2Dt)^{1/2} \tag{3.13}$$

where D is the equilibrium-carrier diffusivity. Such time dependence of the packet spread implies the following dependence of the transit time dispersion Δt_d upon the electric field and the sample thickness,

$$\Delta t_d = \sigma_d(t)/(\mu E) = (2kTL/E\mu^2E^3) \tag{3.14}$$

The Einstein relation between the equilibrium-carrier mobility and diffusivity, $D = kT\mu/e$, was used in the right-hand side of the preceding equation. In the experiments performed by Kasap et al. [11] the dispersion is measured between half-maximum points to represent full width at half maximum and is denoted as Δt_{TOF}.

If charge transport is controlled by multiple trapping of charge carriers (this is typical for noncrystalline semiconductors), the carrier packet also experiences dispersion due to a statistical variation in the trapping and release times; because carriers have spent different times in extended states, they cover different distances. Reasonably, the stronger the external electric field, the larger the typical variation of covered distances, given a fixed variation of trapping and release times. This process was shown to be equivalent to conventional diffusion along the field direction [12–14], with the effective field-associated diffusion coefficient D_f being proportional to the squared field strength as,

$$D_f = \mu_2 E_2 \tau(T) \tag{3.15}$$

where τ is the effective carrier release time determined by the temperature and the density-of-state distribution. It should be noted here that once the equilibrium transport regime is established, the field dependence of the carrier packet spread is no longer sensitive to the density-of-state distribution, although the latter does govern the temperature dependence of the packet width. Because the field-assisted diffusion is equivalent to conventional diffusion and these two processes occur simultaneously, the total packet rms (root mean square) time dispersion σ_T is controlled by a sum of the two diffusion coefficients as,

$$\sigma_T(t) = [(2D + D_f)t]^{1/2} \tag{3.16}$$

The total dispersion observed in the time domain Δt_T is then given by

$$\Delta t_T = [2kTL/(e\mu^2E^3 + 2L\tau/\mu E)] \tag{3.17}$$

The preceding equation shows that the transit time dispersion under weak field conditions is controlled by conventional diffusion, whereas at strong fields, the main contribution to Δt_C arises from the field-assisted diffusion term. A crossover from $\Delta t_T \sim E^{-3/2}$ to $\Delta t_T \sim E^{1/2}$ occurs in the field dependence of the transit time dispersion that corresponds to the crossover from $\Delta t_T \sim E^{-3/2}$ to $\Delta t_T \sim E^{1/2}$ in the dependence of the transit time dispersion on the transit time. It is worth noting that all parameters describing the contribution of the preceding equation are defined by independent measurements, while the contribution of the field-induced diffusion depends on the value of the effective release time, which is poorly known and can be very different in different disordered materials.

All the preceding mechanisms of the carrier packet spread and transit time dispersion imply that charge transport is controlled by traps randomly distributed in both energy and space. This traditional approach completely disregards the occurrence of long-range potential fluctuations. The concept of random potential landscape was used by Tauc [15] and Fritzsche [16] in their models of optical absorption in amorphous semiconductors. The suppressed rate of bimolecular recombination, which is typical for many amorphous materials, can also be explained by a fluctuating potential landscape.

Analytical consideration of the charge-carrier kinetic within a randomly fluctuating potential landscape is virtually impossible, although this concept formed the basis for the model of mobility distribution in amorphous materials. According to the latter model, every carrier photogenerated at the front electrode travels through the sample along its individual route, following mostly valleys and crossing passes within the random potential landscape. This allows us to assume that the drift of every carrier is characterized by its individual average drift velocity and, therefore, by its individual drift mobility.

The function given by $dj(t)/dt$ (calculated case) has a maximum that can be identified as the carrier transit time, which can be derived by using Newton's method of finding the root for the maximum of corresponding equation (for details see Ref. [11]).

Traditionally, charge-carrier transport in pure and doped a-Se is considered within the framework of the multiple-trapping model [17], and the density-of-state distribution in this material was determined from the temperature dependence of the drift mobility and from xerographic residual measurements [18] and posttransient photocurrent analysis.

References

1. B.T. Kolomiets, Phys. Stat. Solidi 7 (1964) 359. 713.
2. J.M. Marshall, Rep. Prog. Phys. 46 (1983) 1235.
3. M.H. Cohen, J. Non-Cryst. Solids 4 (1970) 391.
4. A. Rose, RCA Rev. 12 (1951) 362.
5. N.F. Mott, E.A. Davis, Electronic Processes in Non-Crystalline Materials, second edn., Clarendon, Oxford, 1979.
6. J.R. Haynes, W. Shockley, Phys. Rev. 75 (1949) 691.

7. W.E. Spear, J. Non-Cryst. Solids 1 (1968) 197.
8. G. Pfister, H. Scher, Phys. Rev. B 16 (1977) 2062.
9. M.E. Scharfe, Phys. Rev. B 2 (1970) 5025.
10. J.M. Marshall, A.E. Owen, Phil. Mag. 33 (1976) 457.
11. S.O. Kasap, C. Haugen, B. Polischuk, E.V. Emelianova, A.I. Arkhipov, J. Imag. Sci. Technol. 45 (2001) 30.
12. W.E. Teft, J. Appl. Phys. 38 (1967) 5266.
13. G. Juska, A. Matulionis, J. Viscakas, Phys. Stat. Solidi 33 (1969) 533.
14. A.I. Rudenko, V.I. Arkhipov, Philos. Mag. B 46 (1982) 177.
15. J. Tauc, Mater. Res. Bull. 5 (1970) 721.
16. H. Fritzsche, J. Non-Cryst. Solids 6 (1971) 49.
17. S.O. KasapA.S. Diamond, D.S. Weiss (Eds.), Handbook of Imaging Materials, second edn., Marcel Dekker, Inc., New York/Basel, 2002, p. 329.
18. M. Abkowitz, R.C. Enck, Phys. Rev. B 25 (1982) 2567.

4 Time-of-Flight Experiments in Amorphous Chalcogenide Semiconductors

Contents

There is a great interest in studying the transport mechanism in high-resistivity semiconductors. Many of these studies involve measurements consisting of injecting a thin sheet of charge via light generation or electron beam excitation into the sample near one surface and observing the transport of the sheet through the bulk under the action of an externally applied electric field. This method of experimentation is widely known as the TOF technique.

In recent years, a less familiar technique has been found very useful as a nondestructive method of investigating transport properties of photoreceptors (these are used in nonimpact printers for electronically processed or stored information). The technique is called xerographic TOF (XTOF) and it can be conveniently employed in parallel with the conventional measurements for photoreceptor characterization. In this chapter, we will briefly consider both traditional and xerographic TOF.

The charge transport in amorphous selenium (a-Se) and Se-based alloys has been the subject of much interest and research inasmuch as it produces charge-carrier drift mobility and the trapping time (or lifetime) $\mu\tau$, usually termed as the range of the carriers, which determine the xerographic performance of a photoreceptor. The nature of charge transport in a-Se alloys has been extensively studied by the TOF transient photoconductivity technique (see, for example, Refs. [1–5] and references cited). This technique currently attracts considerable scientific interest when researchers try to perform such experiments on high-resistivity solids, particularly on commercially important amorphous semiconductors such as a-Si and on a variety of other materials

Trap Level Spectroscopy in Amorphous Semiconductors. DOI: 10.1016/B978-0-12-384715-7.00004-8

(including a-Se as a promising material for X-ray digital imaging). The early work involving drift mobility was pioneered by Spear [2]. He and a number of authors successfully applied the technique to solids in the 1950s. Some of important aspects and their interpretation, as well as the principles, have been reviewed by a number of authors [1–5]. It is necessary to note that nearly all TOF (including recent) experiments have been based on the conventional principle of applying a bias voltage, injecting charge carriers by photoexcitation and, finally, by measuring the resulting transient photocurrent that is generated in an external electrical circuit (see Fig. 4.1).

The TOF principle is based on generating a thin sheet of electron–hole pairs near the surface of the sample—usually a high-resistivity semiconductor with negligible thermal equilibrium-carrier concentration, sandwiched between two electrodes, one of which is semitransparent, as shown in Fig. 4.1.

The sample is sandwiched between two electrodes, typically Au and aluminum, the former being reasonably semitransparent in the case of optical excitation. The principle of the technique is illustrated in Fig. 4.1. The wavelength of the exciting radiation is chosen so that the absorption depth δ is much less than the thickness L of the sample. The photogenerated electrons are neutralized almost immediately by reaching the top electrode, but the holes have to traverse the sample toward the grounded (bottom) electrode. As the holes are driven across the specimen by the applied electric field F, they generate an external circuit current $i(t)$, which can be detected via the small resistance R in the circuit. Provided the time constant associated with R and the sample capacitance C_s is much less than the transit time, and if $t_T = L/v_d$ is the drift velocity, the detected signal will be proportional to the photocurrent due to the sweep of the injected holes. It is necessary to note that the experiment in Fig. 4.1 is the same technique as

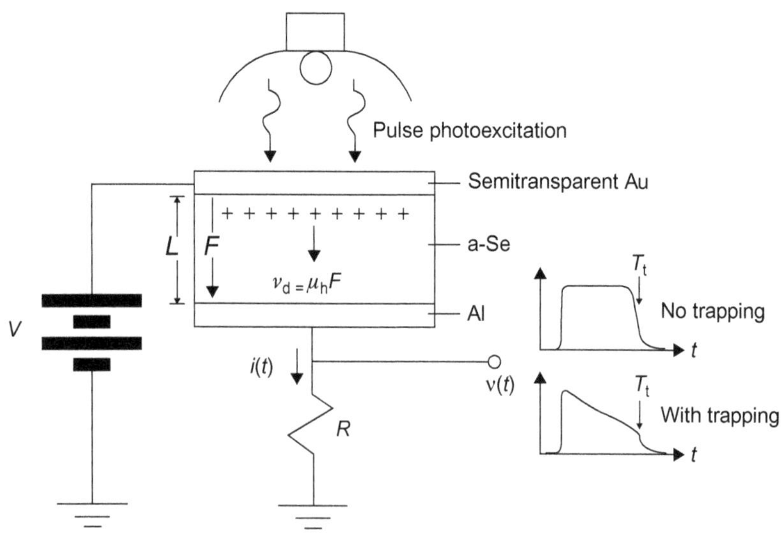

Figure 4.1 Illustration of the principle of the conventional TOF transient photoconductivity technique.

that used [1] for the detection of light pulses by a reverse biased p–i–n photodiode in light-wave communications. Indeed, the basic theoretical framework in both cases for relating the transient photocurrent $i(t)$ to the drift of the injected charge carriers is the p–i–n detector drift in a depleted region. The expected photocurrent transient response from p–i–n photodetectors has been well described in communications engineering in terms of the extent of absorption into the device and the drift of the photogenerated electrons and holes in the depletion region under field-dependent drift mobility or near saturation velocities. In the literature, one can find a wealth of interpretation of photocurrent waveform in terms of the interaction of the photoinjected carriers with various traps in the sample. For example, in the case of amorphous semiconductors, the TOF photocurrents provide a means of obtaining the energy distribution of the localized states in the mobility gap (see, for example, Ref. [5]).

The TOF experiment suffers from the special sample preparation and the potential influence of the electrode. As we will show, these difficulties do not arise in XTOF.

There are several variations to the conventional TOF in Fig. 4.1 in the literature (see Ref. [1]), which frequently involve the following:

a. delayed bias or advanced photoexcitation, for studying recombination [6];
b. delayed photogeneration for studying bulk space charge evolution [7];
c. double photoexcitation for recombination studies [4];
d. interrupted-field TOF (IFTOF) technique, in which the field is temporarily interrupted during the drift of the injected charge carriers [8].

The IFTOF measurement, in essence, involves [1] interrupting the transit of photoinjected charge carriers during their flight across the specimen in the conventional TOF measurement. The interruption is achieved by removing the applied field at time t_1. After an interruption period of t_i, the field is reapplied at time $t_2 = t_1 + t_i$ and the photocurrent is redetected. The resulting nearly-trap-free photocurrent is shown in Fig. 4.2(a) for holes and electrons, respectively at right and left-hand side. Fig. 4.2(b) displays the TOF photocurrent in 18.4 wt% Te:Se, for holes and electrons, respectively (for details see the caption to Fig. 4.2).

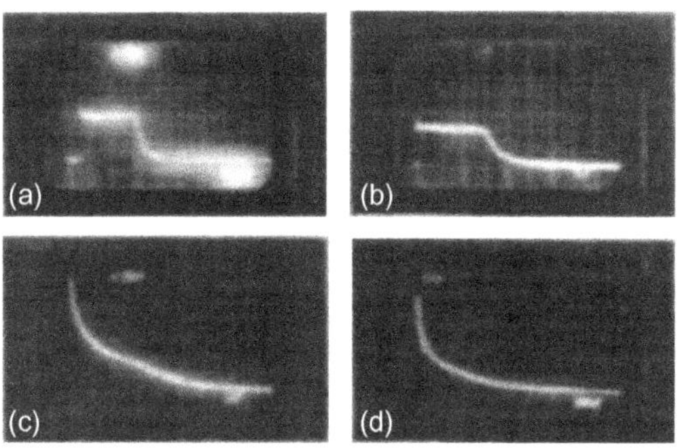

Figure 4.2 Comparison of room temperature transit pulse-shapes for holes on left-hand side and electrons on right-hand side: (a) a-Se, $E = 10\,V/\mu m$, $0.1\,\mu s$/div for holes and $2\,\mu s$/div for electrons; (b) 18.4 wt% Te:Se, $E = 17.5\,V/\mu m$, $1\,\mu s$/div for holes and $200\,\mu s$/div for electrons.

Fig. 4.2(c) displays the interruption of the bias voltage at time t_1 until time t_2, when it is reapplied. The ideal interrupted TOF photocurrent in which displacement currents are absent is displayed in Fig. 4.2(d) as a result of the application of the bias in Fig. 4.2(c). Figure 4.2(a) and (b) also define the various time marks used in the remainder of the extensive article [1].

The same article first critically reviewed the IFTOF principle with its various distinct advantages and then applied the technique to the measurement of the spatial dependence of the hole lifetimes in Cl-doped amorphous Se:0.3%As X-ray plates used in X-ray imaging. The hole lifetime could be measured as a function of location in the term, and the changes in the spatial variation of the lifetime could be determined upon exposure to X-rays. The IFTOF technique is shown to be an extremely powerful tool for studying spatial dependence of charge transport and trapping parameters in the sample.

4.1 An Apparatus for IFTOF Measurements

The IFTOF measurements were carried out by using the experimental setup shown in Fig. 4.3, which is similar to that described previously by Kasap et al. [9] in that it uses a Schering-type bridge to eliminate the displacement of current effects. C_{NS} and C_{N1} are nulling capacitances that are used to balance the bridge to eliminate the displacement current-induced transient voltage between G and H (see Fig. 4.4). The advantage of a Schering-type bridge is that it allows to full applied voltage to appear across the device. Other bridge configurations are also possible [7, 9]. It is essential that the bridge technique for eliminating the displacement current-induced voltage transients only works when the dielectric constant of the sample shows no frequency dependence up to frequencies much greater than $1/t_T$.

If the material exhibits slow polarization currents during switching, then the bridge technique fails, unless one uses an identical sample for the nulling capacitance and employs a resistance ratio bridge.

With the current apparatus, using a totem-pole configuration of totem-pole metal-oxide-semiconductor (TMOS) transistors for switching, bias voltages up to 1 kV could be interrupted so that the a-Se samples of xerographic thickness could be readily examined.

Recently available high-voltage TMOS transistors (MTPIN100) were used in a totem-pole configuration to switch the applied bias for the IFTOF apparatus (see Fig. 4.5). High-voltage TMOS transistors have been successfully utilized in a number of high-voltage, fast switching applications, such as a 1-kV electro-optic driver [10]. The required high-voltage DC source for the switch was provided by a high-voltage Eveready dry cell battery. Simple changing of the high-voltage battery allowed different voltages to be obtained for IFTOF measurements. Because the maximum allowable gate-source voltage for these transistors was ± 20 V, the required gate-source voltage for switching each TMOS was applied from a complimentary-symmetry metal-oxide semiconductor (CMOS) optocoupler, which was powered from a floating 9-V battery. In IFTOF analysis, the switching time of the applied bias must be much shorter than the transit time t_T of the carriers across the sample so that the carriers are either drifting under the presence of the full applied field, or the drift is halted because the applied field

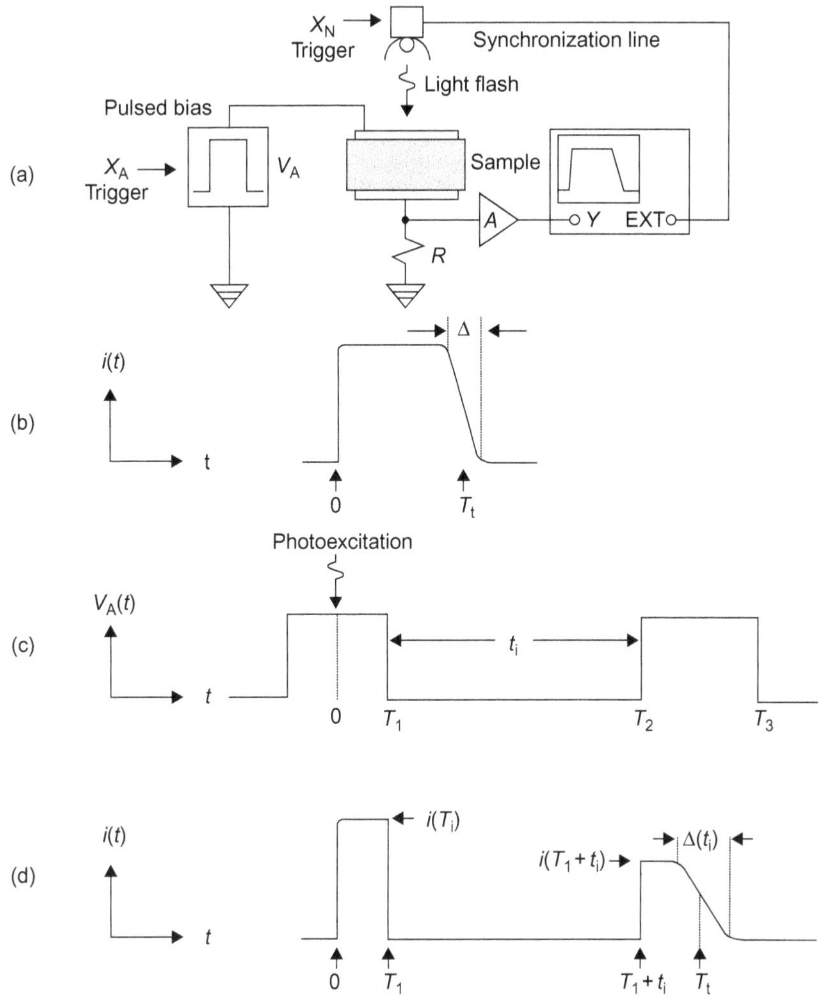

Figure 4.3 Illustration of the IFTOF technique. (a) The conventional TOF measurement technique. V_A is the applied voltage, and X_A and X_N are the triggers for the applied voltage and the xenon flash, respectively. (b) Conventional TOF photocurrent signal. Δ is due to the dispersion broadening of the photoinjected charge sheet; T_t is the transit time. (c) The IFTOF applied voltage profile, $V_A(t)$. (d) IFTOF photocurrent signal resulting from the application of the bias $V_A(t)$ in (c). T_t is the mean time of arrival of the photoinjected holes, which also includes the interruption time t_i [8, 9].

has been completely removed. To speed up the switching time of transistors, common-emitter push–pull driver circuits were added to drive the TMOS gates. Without the push–pull driver circuits, the switching time was approximately 500 ns, which was limited essentially by the output current sink and source capabilities of the CMOS opto-couplers. However, with the driver circuits included, the output of the TMOS switch could swing up to 1 kV in approximately 150 ns [11].

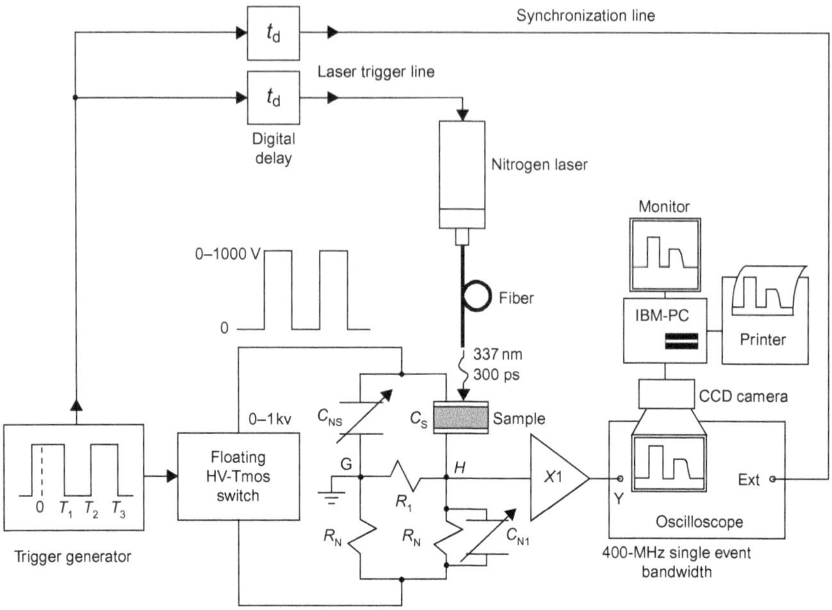

Figure 4.4 Schematic diagram of the high-voltage IFTOF apparatus [1].

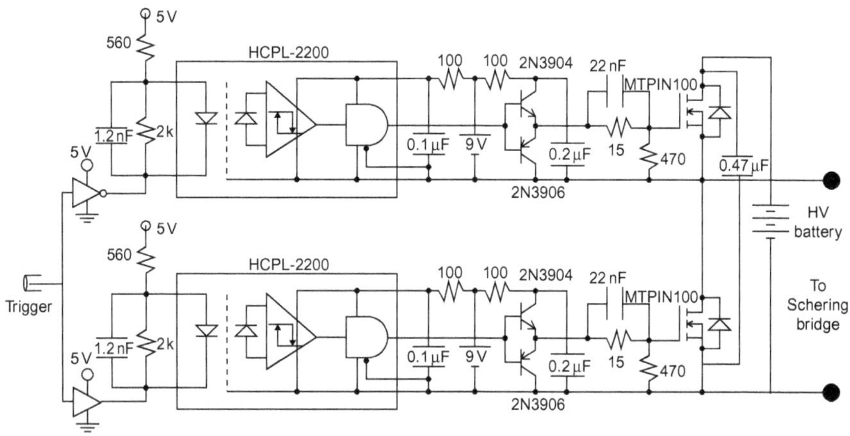

Figure 4.5 High-voltage TMOS voltage switch, which can switch up to 1 kV in a time of 150 ns (from Ref. [1]).

Photoexcitation was achieved by using a triggerable nitrogen gas laser (LGI-21). The laser output pulse was of duration 8 ns at a wavelength of 337 nm.

The IFTOF technique requires the current mode of operation in which the various Resistance-capacity product (RC) time constants in the bridge are much smaller than the transit time t_T. If the resistance R_1 is larger than R_N, then the signal across the amplifier will be simply $R_N i(t)$, where $i(t)$ is the transient photocurrent.

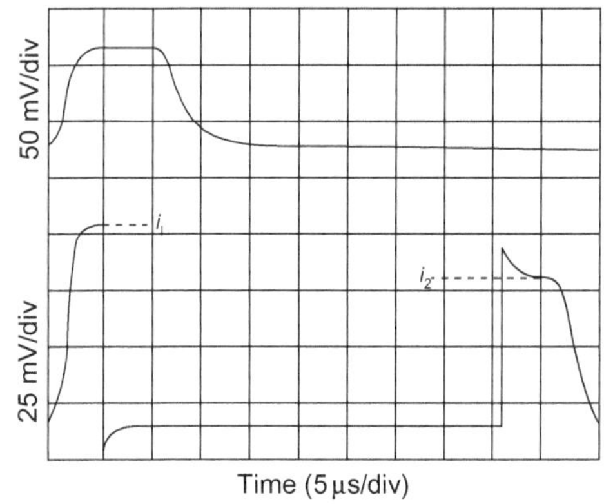

Figure 4.6 Upper trace: conventional TOF photocurrent at a bias voltage of 1 kV. Lower trace: the preceding photocurrent is interrupted for a duration of 40 μs. The interruption voltage is the same, whereas the photosignal is only 80 mV. The fractional-recovered photocurrent is $i_2/i_1 = i(T_1 + t_1)/i(T_1)$.

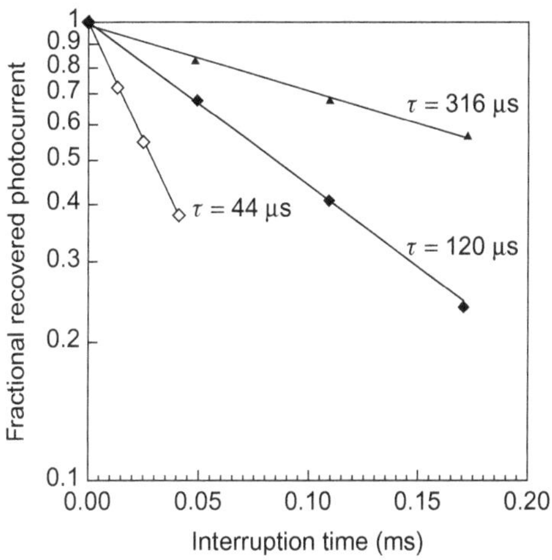

Figure 4.7 Semilogarithmic plot of the fractional-recovered photocurrent versus interruption time. Open and solid diamonds are pure a-Se films, whereas solid triangles are chlorinated a-Se:0.35%As film.

In the current apparatus, the TOF signal was displayed on a 400-MHz analog oscilloscope. The waveform from the screen was coupled into an IBM personal computer for data analysis and storage. In general, TOF experiments are carried out under single-shot measurements, which means that the single-event bandwidth of the oscilloscope must be much larger than $1/t_T$.

Figure 4.6 shows a typical IFTOF photocurrent waveform obtained by interrupting the conventional TOF photocurrent in a Cl-doped a-Se:0.35%As film at 1 kV.

It can be seen that the 1 kV transient switching voltages, which would normally occur at the switching on and off times of the bias voltage, were eliminated down to level where they do not interfere with the photocurrent measurement. Figure 4.7

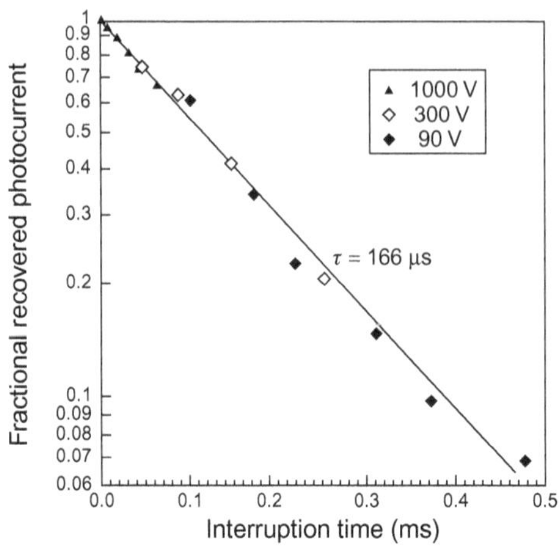

Figure 4.8 Semilogarithmic plot of the fractional-recovered photocurrent versus interruption time at various applied (interruption) voltages.

displays a semilogarithmic plot of the fractional-recovered photocurrent against the interruption time t_i for three typical a-Se and Cl-doped a-Se:0.35%As samples. It was observed that provided the films were bulk-space-charge free as a result of prolonged dark resting between each measurement, the deep-trapping kinetics followed an exponential decrease with a well-defined hole lifetime τ.

No significant difference was observed in the determination of τ if the fractional collected charge (integration of the photocurrent) was examined as a function of t_i instead of the fractional-recovered photocurrent. Furthermore, to confirm that the observed exponential decrease in the fractional-recovered photocurrent is due to a well-defined deep-trapping time τ, IFTOF measurements were also carried by interrupting different bias voltages and by interrupting the drift of the photoinjected carriers at different locations in the specimen as shown in Figs. 4.8 and 4.9. As expected, τ manifested no dependence on the applied field, indicating that during the interruption period t_i, charge-carrier trapping is essentially occurring under a negligible internal field (in other words, due to the charge packet itself). This situation is summarized in Fig. 4.8 by a series of experiments. As indicated by the next series of experiments (see Fig. 4.9), the trapping signature $\log(i_2/i_1)$ versus t_i at different locations in the sample was obtained by interrupting the photocurrent at different times before the transit time. Furthermore, by using semitransparent electrodes on both sides of the sample, photogeneration could be achieved at either surface, or the drift of holes to or from the substrate could be studied. The trapping signatures showed no dependence on the location of interruption or on the direction of propagation. Thus, it may be assumed that the hole lifetime is relatively uniform across the sample. It should be emphasized that Figs. 4.8 and 4.9 demonstrate quite clearly the power of the IFTOF technique for studying charge trapping in high-resistivity semiconductors.

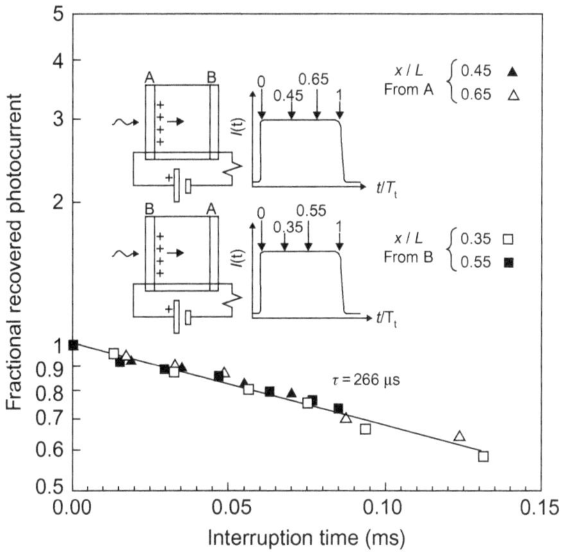

Figure 4.9 Semilogarithmic plot of the fractional-recovered photocurrent $i(T_i + t_i)/i(T_i)$ versus interruption time t_i at two different locations in the specimen. Interruption locations are indicated on the schematic TOF waveforms in the upper inset. Interruption at location $(x/L) = 0.65$ when photoexcited on side A corresponds to location $(x/L) = 0.35$ when photoexcited from side B.

4.2 XTOF Technique

As mentioned in the previous paragraphs and in Chapter 3, there is a great interest in studying the transport mechanism in high-resistivity semiconductors. Usually these measurements consist of injecting a thin sheet of charge via light generation or electron beam excitation into the sample near one surface and observing the transport of the sheet through the bulk under the action of an externally applied electric field.

In recent years, with the emergence of nonimpact printers for electronically processed or stored information, a less familiar technique has been found very useful. The technique is a nondestructive method of investigating transport properties of photoreceptors that are used in these systems. The technique is called XTOF, and it can be conveniently employed in parallel with the conventional xerographic measurements for photoreceptor characterization. Here we note that the TOF experiments suffer from the special sample preparation needed and the potential influence of the electrode. These difficulties do not arise in XTOF.

The purpose of this section is to introduce a practical XTOF system that can be simply and inexpensively constructed to carry out transport measurements on high-resistivity semiconductors.

XTOF was successfully carried out at the IBM research laboratories [12] and more recently at the Xerox laboratories [13]. The principle of operation of the techniques has been described by these authors. The theory involved in the interpretation of the XTOF transient signals is the same as that for TOF signals, which has been reviewed by a number of authors [13, 14].

Figure 4.10 shows a simplified XTOF experiment and clearly demonstrate the principles involved.

In the following, we consider the principles of the XTOF technique. In these experiments, the floating surface of the sample rests on an earthed electrode and is charged in the dark by a corona charging device. A short pulse of strongly absorbed light is used to generate a thin sheet of electron–hole pairs near the surface of the sample. As in the case of conventional TOF, the wavelength of the light is chosen so that the absorption depth is $\delta(\lambda) \ll L$ (L is the thickness of the sample). As is clear from Fig. 4.10, the electrons neutralize (in the chronological order) some of the positive charges on the surface, whereas holes move toward the conductive substrate under the influence of the applied field.

Usually, the number of injected carriers N_t is much smaller than that of the surface charge. Thus, the drifting sheet of charges has a negligible effect on the applied field (the internal field can be taken as $E_0 = V_0/L$, where V_0 is the sample voltage). This is a so-called small-signal operation mode and was widely used in theoretical analysis of the transient photocurrent [16]. The injected charge sheet moves under the action of the applied field E_0 with a constant drift velocity $V_d = \mu_d F_0$, where μ_d is the drift mobility. The time taken for the charge sheet to reach the substrate (bottom electrode), where it is neutralized, is called the transit time t_T,

$$t_T = L^2/\mu_d F_0 \tag{4.1}$$

The drifting charge sheet ($t \leq t_T$) is equivalent to a transient current $I(t)$. The latter depends on the drift velocity V_d of the charge sheet and the total charge qN_t in it.

$$I(t) = qN_t V_0 \tag{4.2}$$

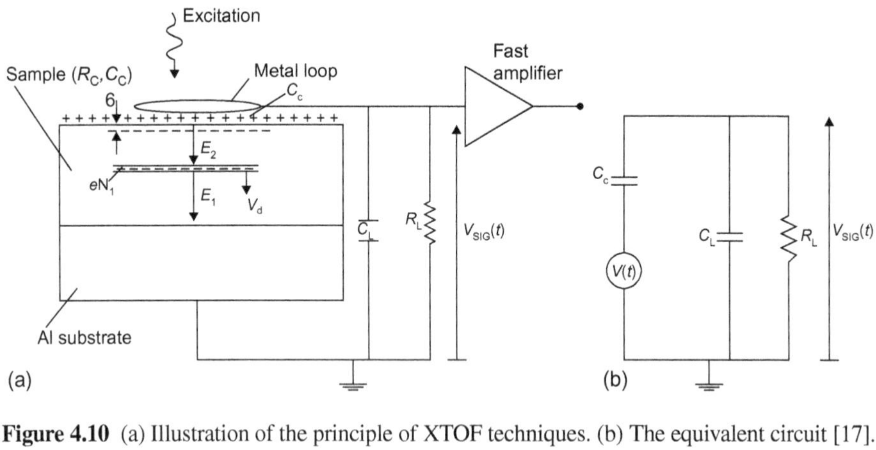

Figure 4.10 (a) Illustration of the principle of XTOF techniques. (b) The equivalent circuit [17].

For time $t > t_T$, this current is zero because the carrier would have reached the substrate. The current $I(t)$ can be detected via the voltage it induces in the external circuit. Shortly, the equivalent electrical circuit of the XTOF experiment under the small-signal condition contains the coupling capacitance C_L (the sum of the amplifier and the parasitic capacitances) and R_L (R_L is the load resistance). The total current is the sum of the conduction current due to the drift of photogenerated charge and the displacement current and is equal to zero (for further details, see Ref. [15]),

$$j_c(t) + \varepsilon_0 \varepsilon_s \frac{dE}{dt} = 0 \tag{4.3}$$

Here, the first component is the conduction current and the rest of the symbols have their usual meaning. Equation (4.3) can also be written as

$$\frac{dV}{dt} = \frac{-L}{\varepsilon_0 \varepsilon_s} J_{c(t)} \tag{4.4}$$

The Laplace transform of signal voltage is given by

$$V_{sig}(s) = \frac{sR_L C_C V(s)}{1 + sR_L(C_C + C_L)} \tag{4.5}$$

Because $C_C \ll C_L$, equation (4.5) can be written as

$$V_{sig}(s) = \frac{sR_L C_C}{1 + sR_L C_C} V(s) \tag{4.6}$$

Equation (4.6) has two convenient solutions for $0 \leq t \leq t_T$, which arises from conditions $R_L C_L \ll t_T$ and $R_L C_L \gg t_T$:

$$V_{sig}(s) = sR_L C_s V(t) \tag{4.7}$$

and

$$V_{sig}(t) = R_L C_s \, dV/dt \tag{4.8}$$

respectively.

Substituting dV/dt (see equation (4.4)), one can obtain

$$V_{sig}(t) = \frac{R_L C_s L}{\varepsilon_0 \varepsilon_s} J_s(t) \tag{4.9}$$

or

$$V_{\text{sig}}(s) = \frac{C_s}{C_L} V(s) \tag{4.10}$$

The next step is to substitute $V(t)$ from equation (4.4):

$$V_{\text{sig}}(t) = \frac{-C_s L}{C_L \varepsilon_0 l \varepsilon_s} \int_0^{t_T} J_s(t)\mathrm{d}t \tag{4.11}$$

Summing up, we have essentially two modes of operation, which depend on the magnitude of the time constant $R_L C_L$. For small $R_L C_L$, the signal is proportional to the conduction current (equation (4.9)) and the so-called current mode. There is another mode of operation—called voltage mode—when $R_L C_L$ is large and the signal is proportional to the integral of the current (equation (4.11)).

Figure 4.11(a) shows the expected signals for two modes of operation when the drifting charge sheet suffers no loss of carriers arising from trapping or recombination.

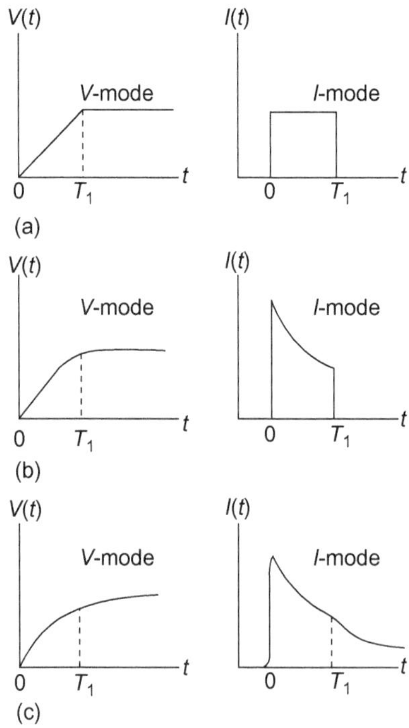

Figure 4.11 Transient current and voltage waveforms showing effect of trapping and detrapping. (a) No trapping. (b) Trapping, no detrapping. (c) Trapping and detrapping (from Ref. [13]).

If some of the carriers are removed from the transmitting charge sheet by trapping, $N_t(t)$ in equation (4.1) exhibits a simple exponential decay, which may be expressed by the following equation:

$$N_t(t) = N_0 \exp(-t/\tau) \tag{4.12}$$

where τ is lifetime and N_0 is the number of carriers created at $t = 0$. Thus, with trapping the current becomes

$$I(t) = (qN_0/t_T)\exp(-t/\tau) \qquad (t < t_T) \tag{4.13}$$

Figure 4.11(b) shows transient current in the presence of trapping for two modes of operation. In this case, the transient current waveforms will have two components corresponding to the parts of the waveforms: $t > t_T$ and $t < t_T$. These are called the fast and slow components, respectively, and they are shown in Fig. 4.11(c). The first component is associated with the drifting of the charge sheet, and the slow component, which has a longer time constant, corresponds to one or more trapping and detrapping events occurring during the transit. We note here that the $V(t)$ waveform is continuous, whereas the $I(t)$ curve has a break at t_T.

The transit time t_T is more clearly defined by the break in the current signal. Thus, it is preferable to operate the apparatus in the I-mode. Figure 4.12 shows transient

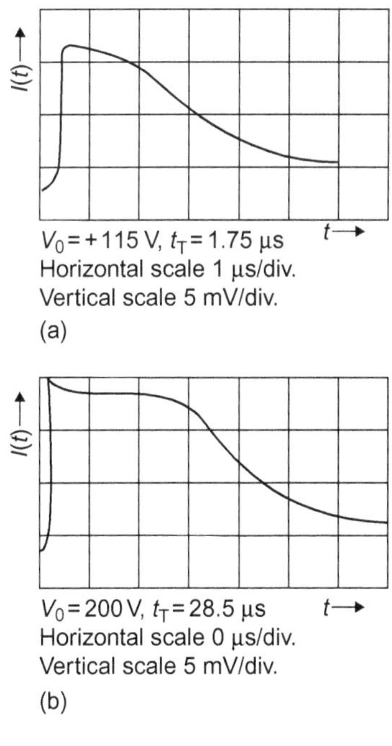

Figure 4.12 Oscilloscope traces of typical XTOF hole photocurrent signals for (a) pure Se sample of thickness 58 µm and (b) 20-ppm Cl-doped a-Se + 5.3% Te of thickness 58 µm.

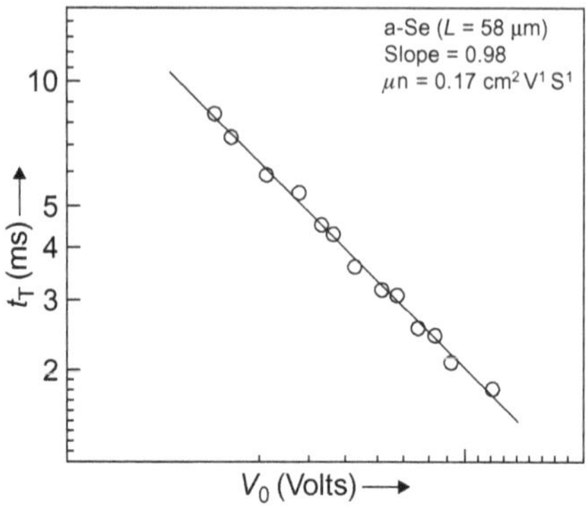

Figure 4.13 Log–log plot of XTOF hole transit time t_T versus applied voltage V_0 for an a-Se sample [13].

current signals observed in pure a-Se and 20-ppm Cl-doped a-Se + 5.3% Te for hole transport under the *I*-mode of operation.

The hole drift mobility can be conveniently obtained from the slope of the curve t_T versus V_0 (see Fig. 4.13) [15].

4.3 TOF Measurements in Selenium-Based Amorphous Multilayer Photoconductors

In recent years, there has been a lot of interest in amorphous selenium-based multilayer photoreceptors for xerography and laser printing [17–21]. In these devices, the photogeneration of charge and subsequent transport of carriers is separated functionally, i.e., these processes occur in different layers. Multilayer structures are very advantageous because the composition of different layers can be tailored independently to meet specific requirements such as spectral sensitivity, good charge transport properties, stabilization, and resistance against environmental interactions. A typical double-layer photoreceptor structure consists of a relatively thin (a few micrometers) Se-based layer for carrier generation, referred to as the photogeneration layer (PGL), and a thick (≈50 μm) pure selenium layer, referred to as the charge transport layer (CTL). There are also triple-layer photoreceptors consisting of a transport layer, a PGL, and a third thin layer whose function is to protect the PGL and improve the charge acceptance of the photoreceptor.

The current trend in photoreceptor design is to use multilayer structures of various Se-based compositions. In particular, certain additives, e.g., As, Te, and Bi, have been shown to extend the spectral sensitivity of selenium to longer wavelengths [17, 20, 21]. Arsenic has been particularly effective in providing increased thermal stability and resistance to crystallization.

In characterizing the electrical properties of Se-based photoreceptors, the TOF technique [22, 23] has been proven to be a powerful tool in basic investigations, as well as for diagnostic purposes. Typical information obtained from TOF studies includes carrier mobility and number and/or depth of traps in the photoreceptor. The interpretation of TOF signals obtained from monolayer photoconductors with nondispersive transport is relatively simple and straightforward. In the case of As–Se and Se–Te alloys, the carriers—which were initially localized in a narrow sheet—spread out due to trapping as they traverse across the photoconductor. This causes a significant distortion of the TOF signal, which makes the evaluation much more difficult. Both the electron and hole drift mobility have been studied in different compositions of amorphous As_xSe_{1-x} alloys (for details, the reader is referred to the review articles [24, 25]), but the investigations were restricted mostly to lower As concentrations. The shapes of transients have not been analyzed in more detail. Thus, we discussed the evaluation of drift mobility of charge carriers in multilayer structures and their dependence on the composition. The data show that the usage of a double-layer structure is the enabling tool to obtain information on dispersive transport. The reasons for choosing a-Se for such a study in this case are the following. First, the properties of a-Se are well documented [18, 20, 22–28]; as such, it can serve as an ideal test and model material for the comparison of various results on mono- and multilayer structures. Second, the exact nature of traps (both shallow and deep) in a-Se and in its technologically important alloyed forms such as As_xSe_{1-x} have not been conclusively determined. A further reason for using a-Se is that it can be readily prepared by using conventional vacuum-deposition techniques with reproducible properties so that the results presented herein will be typical for any pure or alloyed a-Se film.

Single- and double-layer xerographic photoreceptors were fabricated by vacuum evaporation of vitreous pellets from stainless steel boats onto SnO_2-coated glass (or oxidized aluminum) substrates held at room temperature. Bulk As_xSe_{1-x} glasses were prepared by a conventional melt-quenching method. To prepare a particular composition, appropriate quantities of high-purity constituent elements sealed in evacuated quartz ampoules were rotated continuously to ensure thorough mixing of the constituents. The ampoules were then quenched in ice water.

The coating chamber was equipped with a set of independently controlled stainless steel boats and a shutter system to enable the fabrication of multilayer structures. Pure selenium pellets were loaded into one boat and As_xSe_{1-x} alloys into another. The two sources were evaporated sequentially (without breaking the vacuum) at boat temperatures of about 450 K. Typical coating rates were 1 μm/min. After evaporation, they were allowed to anneal over several weeks in the dark at room temperature. During this period, due to structural bulk relaxation, most physical properties of the photoconductor film become stabilized. The compositions of the deposited films were determined by electron probe microanalysis, and the compositions quoted ($0 \leq x \leq 0.20$) are accurate to within 0.5 at.%. By shuttering the beginning and the end of the evaporation, a uniform arsenic composition across the film thickness can be obtained. In all experiments, a transparent gold electrode (~300 μm thick) was used as the top contact.

Electroded TOF experiments have already been described previously [29]. They were carried out under small-signal conditions to determine hole and electron drift mobility. Carriers were generated by illuminating the sample through the semitransparent Au electrode with a pulsed N_2 laser at 337 nm. The duration of the pulsed light was about 10 ns, which was sufficiently shorter than the carrier transit time. Because the photocarriers are generated at the sample surface due to the light absorption coefficient of As_xSe_{1-x} glasses at 337 nm ($>10^4\,cm^{-1}$), the species of drifting carriers can be chosen by changing the polarity of applied voltage across the sample. The transient current was amplified and displayed on a storage oscilloscope. The transport of carriers in the corresponding layers was evaluated by time-resolving the transit signal. To avoid cumulative bulk space charge buildup, after each excitation, the sample was discharged by successive nitrogen laser pulses and then allowed to rest in the dark for several minutes (so-called single shot mode of operation).

4.3.1 Monolayer Systems

Figure 4.14(a) shows a typical TOF electron signal of a space-charge-free amorphous Se monolayer. We see that the transient profile for a well-rested (dark adapted) is a quasi-rectangular pulse. The photocurrent remains constant up to 1 μs, and then decreases abruptly. The signal indicates essentially nondispersive transport at room temperature. When the carriers arrive at the substrate, the signal decreases quickly to zero. This is the transit time t_T, which can be readily determined from the break point. The drift mobility is then calculated from $\mu^e = d^2/(t_T V_a)$, where d is the sample thickness and V_a is the applied voltage. This room temperature value of the drift mobility, $\mu^e \approx 5 \times 10^{-3}\,cm^2/Vs$, is in remarkably good agreement with reported values [24, 30].

Addition of As to Se causes the transport to become more dispersive. Figure 4.14(b) shows the corresponding transient characteristic of a single-layer As_xSe_{1-x} structure.

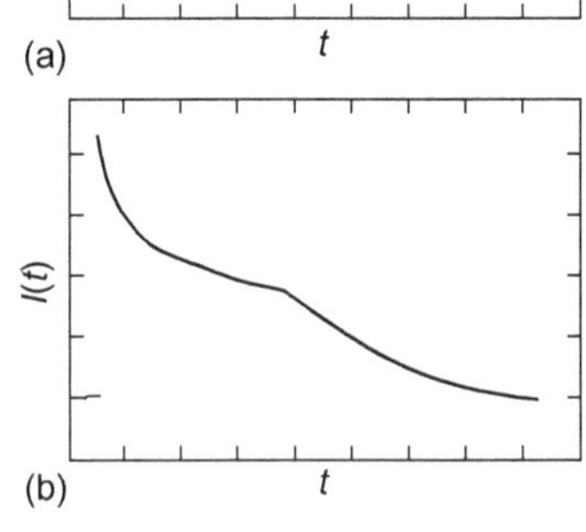

Figure 4.14 Typical TOF traces for monolayer structures. (a) Pure Se, 10 mA/div, 0.2 μs/div, 9×10^4 V/cm, $d = 3.2$ μm. (b) $As_{0.08}Se_{0.92}$, 5 μA/div, 20 μs/div, 1.4×10^5 V/cm, $d = 3.3$ μm.

The signal contained an initial decay in the form of a spike. We may consider three possibilities as originally causing the photocurrent decay in the pretransit region.

First, because the method of excitation involves the creation of a charge sheet near the top surface of the sample, it has been argued [31] that the initial spike is due to the movement of the holes toward the top surface. If this is the case, then, due to the larger difference in electron and hole mobility in our samples, a large spike due to holes in the electron response should be observed. Furthermore, the duration of the spike should be inversely proportional to the appropriate mobility. Second, the current spike could be caused by a high-field region in the vicinity of the illuminated electrode. Next, the initial current decay can be understood in terms of relaxation of charge photoinjected into a distribution of localized states. A detailed analysis of transient signals in the As_xSe_{1-x} system supports this third explanation for the spike. For samples As_xSe_{1-x} with $x < 0.10$ at any fields applied in this experiment, a shoulder is always clearly revealed in the current decays, followed by a long tail; the transit time is still discernible. As the arsenic concentration increases further, the pronounced break point becomes progressively more difficult to identify. Transit pulse shapes have become highly dispersive, meaning that there is now a broad statistical distribution in photoinjected-carrier transit times. In these alloys, photoinjected-carrier transport is generally interpreted [20, 24] as shallow trap-controlled transport, which involves multiple trapping of transiting carriers from the band into shallow states. Thus, the increasing dispersiveness of TOF transients can be understood in terms of a composition-induced broadening in the respective interactive shallow trap manifolds.

4.3.2 Multilayer Systems

The transients corresponding to carrier movement across multilayer samples have different profiles from monolayer samples. A steplike transient current waveform is expected due to the difference of mobility between the two layers (namely, PGL and CTL). However, as is evident from Fig. 4.15, in actual photoreceptors this is not the case; rounded waveforms are observed. The signal consists of two parts. The initial part represents the charge movement through the As_xSe_{1-x} PGL.

Because of the low mobility in this material, the carriers move with relatively low speed. The magnitude of the signal is small and decreases even further as carriers are trapped in the layer. At time t_0, the fastest carriers arrive at the interface between the PGL and CTL and enter the CTL. There they move with a much higher speed, according to their large mobility in Se, and induce a much larger signal in the measuring circuit. It is necessary to note here that the portion corresponding to transport in a-Se CTL is now distorted in comparison with that observed for the a-Se single layer. The reason for this is the spreading of the propagating carrier packet in the PGL. The signal increases with the increasing number of carriers entering the transport layer. After the transit time through the transport layer, the carriers recombine with the charge at the substrate. The signal becomes zero when the total charge has finished the transit through the transport layer. Thus, the separate contributions from transport in the two layers are clearly in evidence. From the inflection points, transit times t_0 and t_T were measured.

The assignment of the inflection points in the waveform to transit times t_0 and t_T is supported by the following arguments. First, t_0 and t_T scaled correctly with the thickness of PGL and CTL in TOF experiments. Next, from the knowledge of transit times t_0 and t_T, applied voltage V_0, and thickness d_1 and d_2, the drift mobility of charge carriers μ_1 and μ_2 in both layers are calculated from the following equations [32]:

$$\mu_1 \approx \frac{d_1^2(1-\varepsilon_1/\varepsilon_2)+d_1 d\varepsilon_1/\varepsilon_2}{t_0 V_0} \tag{4.14}$$

$$\mu_2 \approx \frac{d_2^2(1-\varepsilon_2/\varepsilon_1)+d_2 d\varepsilon_2/\varepsilon_1}{t_T V_0} \tag{4.15}$$

where t_0 and t_T are transit times for the PGL and CTL, respectively, and $d = d_1 + d_2$ is the total thickness. Note that these data are in general agreement with those obtained for monolayer samples of the same composition and thus support the assignment of t_0 and t_T as transit times. Furthermore, t_0 and t_T are independent of the substrate materials and top contacts.

In order to examine the contact effects on both μ_1 and μ_2, TOF was carried out on similar samples with different top and bottom contacts (gold, aluminum, copper, and SnO_2). From the results, it can be concluded that both μ_1 and μ_2 are independent of the top and bottom contact and also of the applied field. These observations indicate

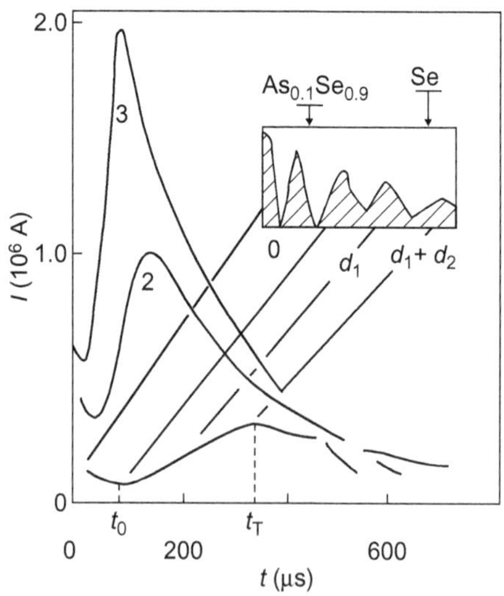

Figure 4.15 Traces of transient electron currents in an a-$As_{0.10}Se_{0.90}$/Se double layer as a function of the layer thickness of $As_{0.10}Se_{0.90}$: (1) $d_1 = 0.9\,\mu m$, (2) $0.7\,\mu m$, and (3) $0.5\,\mu m$. The applied electric field is 7.2×10^4 V/cm. The inset schematically shows the spatial distribution of photocarriers at different times.

clearly that μ_1 and μ_2 are meaningful parameters characteristic of bulk transport in PGL and CTL.

Figure 4.16 shows a series of waveforms of transient photocurrent for various temperatures.

We see that the transit signal does not change appreciably with temperature. Also note that at the final stage ($t > t_T$), the TOF signal exhibits a power-law decay and appears to be dispersive. According to the Scher–Montroll theory [33], in the case of the dispersive transport process, $I(t)$ should exhibit power-law dependences $t^{-(1-\alpha)}$ and $t^{-(1+\alpha)}$ for $t < t_T$ and $t > t_T$, respectively, where α is the disorder parameter. The latter can be determined from both the initial (α_i) and final (α_f) slopes. In our results, the condition $\alpha_i = \alpha_f$ predicted by Scher and Montroll is generally not observed; α_i and α_f have different field and temperature dependencies. The inset in Fig. 4.16 shows the temperature dependence of the drift parameters t_0 and t_T. The Arrhenius plot gives a good straight line in the temperature range 250–300 K, showing an activation-type process of electron transport in the current samples. The activation energy can be obtained from the slope. The values of E_μ in Fig. 4.16 are estimated to be 0.33 and 0.35 eV and are close to those of pure Se and $As_{0.10}Se_{0.90}$, respectively. The preceding results are taken as evidence that electron transport in multilayer structure is still nondispersive.

The triple-layer device shows a transient characteristic with a more complicated signature (Fig. 4.17). The photocurrent can roughly be divided into three time regions. Transport in the thin top layer (a-Se) appears as a narrow pulse. An initial current rises to a peak value and then decays to a smaller value. This is defined as a "spike." There follows a region of low signal amplitude (the "saddle region"), corresponding to transport in the thin, low-mobility middle layer. Finally, the growth of current terminates in a broad rounded peak, which is then followed by a slowly decaying tail. This broad portion of the signal ($t > t_2$) represents transport in the high-mobility bottom layer. As the applied field across the multilayer structure increases, t_1, t_2, and t_3 decrease.

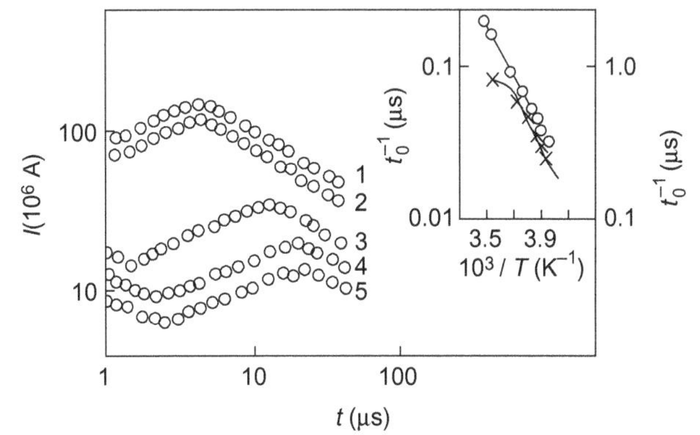

Figure 4.16 Temperature dependence of transient electron currents in an a-$As_{0.10}Se_{0.90}$/Se double layer ($d_1 = 0.3\,\mu m$ and $d_2 = 5\,\mu m$) at 8×10^4 V/cm: (1) $T = 290$ K, (2) 283 K, (3) 270 K, (4) 265 K, and (5) 260 K. In the inset, the inverse of electron transit time t_0 and t_T is plotted against reciprocal temperature.

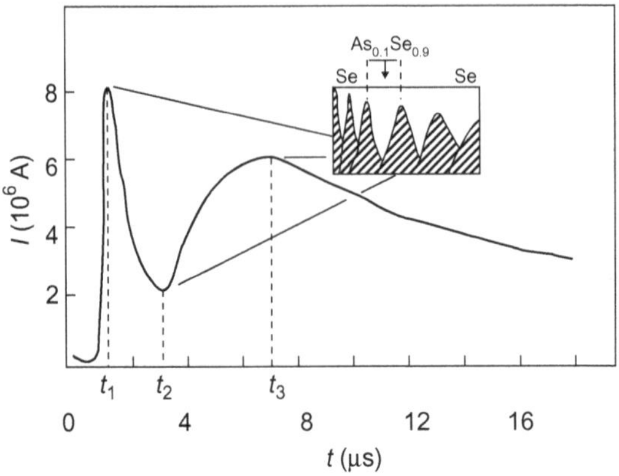

Figure 4.17 A typical transient electron current waveform in an a-Se/$As_{0.10}Se_{0.90}$/Se triple-layer sample. $d_1 = 0.81\,\mu m$, $d_2 = 0.43\,\mu m$, and $d_3 = 8.40\,\mu m$. $E = 8 \times 10^4$ V/cm. The inset shows the spatial distribution of photocarriers at different times.

The assignment of t_1, t_2, and t_3 (see inflection points in Fig. 4.17) to transit times in the top, middle, and bottom layers is supported by the fact that the drift mobility of charge carriers for the three layers were calculated to be similar to the corresponding single layers. The general features of current waveforms described earlier are common to both hole and electron response.

4.3.3 The Effect of Interface

A potential barrier may be formed at the interface between the As_xSe_{1-x} and Se layers due to the difference of optical band gaps. Thus, the transport can be influenced by injection characteristics. We have calculated the carrier injection efficiency As_xSe_{1-x} to a Se layer. For example, for double-layer photoreceptors, this is achieved by calculating the charge that reaches the interface between PGL and CTL and the total charge collected at the end of the transit time. Such calculations can be easily made by integrating the transient currents for each layer. The injection efficiency is independent of the applied electric field in the range between 4×10^4 and 1×10^5 V/cm. These results suggest that the charge accumulation at the interface does not significantly affect the charge transport. The physics of the interface states between different mobility gap amorphous layers have escaped any detailed investigation; as a result, it is not possible to extend the experimental observation here to amorphous semiconductors in general.

4.3.4 Light-Induced Effects on Photocurrent Transients

It is well known that amorphous chalcogenide semiconducting materials exhibit many kinds of photoinduced phenomena (photodarkening and photocrystallization, to mention only a few) for which extensive work is now being carried out because of their fundamental interest [34, 35]. Further, the phenomena have also received attention because of their potential application for optical memories [36]. While there are

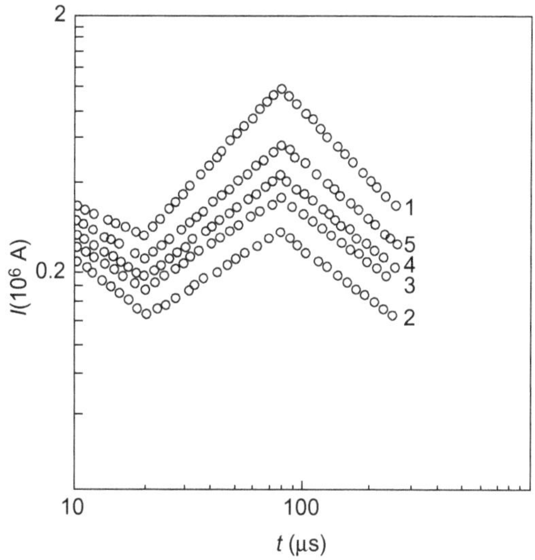

Figure 4.18 Transient electron current in a-$As_{0.10}Se_{0.90}$/Se double-layer samples. $d_1 = 0.5\,\mu m$, $d_2 = 9.6\,\mu m$, $E = 7.2 \times 10^4$ V/cm, $T = 293$ K: curve 1, dark rested; curves 2–5, exposed to band-gap illumination and then dark rested for 5, 20, 30, and 40 min, respectively.

many experimental studies on the change in structural and electronic properties, only a few studies concerning light-induced effects on photocurrent transients have been reported [37–39]. To obtain further insight into the photoinduced electronic effects, for the first time, we have examined transient drift characteristics in amorphous multiple layers that were exposed to band-gap light. Figure 4.18 shows a typical trace of the transient electron in the dark-rested and preexposed $As_{0.10}Se_{0.90}$/Se sample. It is apparent that band-gap irradiation of the sample leads essentially to a decrease of the current level, whereas the transit time t_0 remains unchanged ($t_0 \approx 20\,\mu s$ for the data in Fig. 4.18). If, after preillumination, the sample is allowed to relax in the dark, the previous (dark-rested) magnitude and shape of the transient current is fully recovered. The important result of this experiment is the observation that the electron drift mobility E_μ^e is the same before and after light exposure. Consequently, shallow states that control charge transport and define the activation energy E_μ^e of the mobility should not undergo photoinduced changes. At the same time, it should be stressed that the observed change in the basic xerographic parameters—i.e., the dark discharge rate, initial charging potential, residual potential, and its dark-decay rate—indicate photoinduced changes of deep states with $E_i > E_\mu^e$.

4.3.5 Utilization of Multilayer Structures for the Determination of Transit Time

A primary goal of TOF experiments is to extract basic material parameters that may be used for more general applications. Of particular importance are estimates of transit times and carrier mobility. It is well known that TOF measurements of the

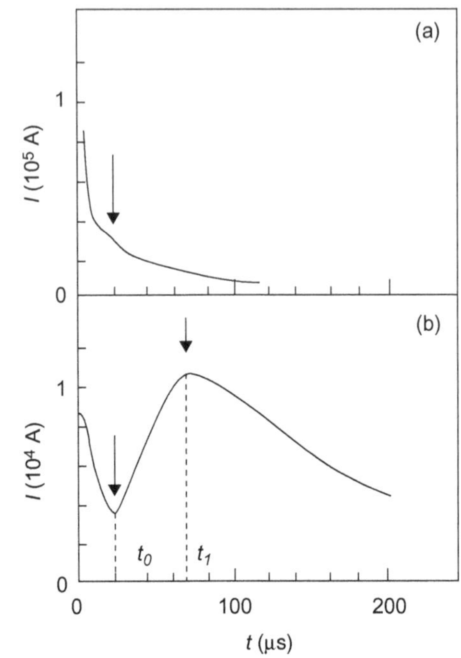

Figure 4.19 Traces of transient hole currents in (a) a-$As_{0.4}Se_{0.6}$ monolayer and (b) a-$As_{0.4}Se_{0.6}$/Se double layer. $d_1 = 1.5\,\mu m$, $d_2 = 25\,\mu m$, $E = 3 \times 10^4\,V/cm$.

drift mobility in As_2Se_3, As_2S_3, and some polymeric films are complicated by a significant spreading of the carrier packet as it propagates through the sample [33, 40, 41]. If carrier transport is very dispersive, a current pulse may appear featureless. However, if the current pulse is displayed using logarithmic axes, a discontinuity of gradient is observed that corresponds to the transit time of the fastest carriers.

In this section, we propose that the transit time of transport photocarriers can be obtained by an analysis of transient current in a double layer that consists of a thin chalcogenide under test and another material with higher mobility such as a-Se. Figure 4.19(a) shows the type of carrier transit pulse that is frequently encountered in the study of amorphous semiconductors.

An anomalously high degree of dispersion is immediately evident from the figure. There is no "flat part," characteristic of idealized current response, to specify a transit time. The transit time is defined by the $\log I - \log t$. It is necessary to note that there exist some disadvantages of using logarithmic conversion of the transient current to extract the transit time. In fact, the intersection of power laws $t^{-(1-\alpha)}$ and $t^{-(1+\alpha)}$ has been used extensively to define the transit time. However, there is a good deal of arbitrariness to this definition. Experimentally, one has access to only limited sections of pre- and posttransit currents.

On the other hand, better time resolution has been obtained in TOF experiments with multilayer structures. In the case of a double-layer configuration (Fig. 4.19(b)), the transit for the $As_{0.4}Se_{0.6}$ layer could be accurately assessed. The waveform clearly showed a discontinuity in the form of an inflection point at $t = t_0$. The time designated with an arrow is used as a transit time. This technique of time-resolved transient current measurements in multilayer structures is especially suited to low-mobility, high-resistivity amorphous chalcogenides with dispersive transport characteristics.

We have considered transient drift characteristics in Se-based multilayer devices using the TOF technique. To illustrate the method of analysis of TOF transient signals, various experimental current waveforms were presented. The shape of the photocurrent waveform indicates two separate transit times for two layers of a double-layer structure. The gradual addition of arsenic progressively decreases μ^e. Principal observations are then an increase in dispersion with a reduction in mobility as As is alloyed with a-Se. It is reasonable, within the shallow trap-controlled mobility model [23, 24], to interpret our observations in the following manner. In a-Se, the electron transport is controlled by a narrow manifold of traps displaced approximately 0.33 eV from the conduction-band mobility edge. The transport tends to become dispersive when kT is less than the nominal trap manifold width δE. It is necessary to note here that in the simulation studies of Marshall [41], various distributions of trapping centers were examined, and it was established that in each case, highly dispersive transit pulses could be generated if the localized states extended more than a few kT. A distribution of the Gaussian form is capable of generating values of α_i and α_f with temperature dependence very similar to that of typical experimental data for a-Se. The addition of As to a-Se, we believe, broadens the distribution of shallow traps, thus shifting the threshold for nondispersive transport to higher temperatures. The decrease of electron mobility with increasing As content could be accounted for by an increase in the density of shallow traps, with E_μ remaining constant at approximately 0.33 eV. Unlike the electron case, the effect of As on hole transport cannot be explained by a gradual change in the density and energy distribution of shallow traps. We can attribute the lifetime limited hole signal in the concentration range 2–6 at.% As to the loss of carriers from the charge packet due to deep trapping. These results are in good agreement with our previous data on monolayers [42] and the data of other investigators [20, 24, 43].

Further, we have also shown that the transient photocurrents in a double-layer structure can conveniently be used for spectroscopic purposes in situations where the use of a conventional TOF technique is limited.

Our experimental results show that irrespective of the exact nature of interface, transport of charge carriers through double-layer (Se-rich) samples does not experience significant deep trapping.

Finally, more details about TOF experiments in various chalcogenide amorphous semiconductors may be found in the articles of the authors listed at the end of this chapter [44–53].

References

1. B. Polischuk, S.O. Kasap, V. Aiyah, A. Baillie, Int. J. Electron. 76 (1994) 1029. see also B. Polischuk, S.O. Kasap, V. Aiyah, A. Baillie, M.A. Abkowitz, Can. J. Phys. 69 (1991) 361.

2. W.E. Spear, Proc. Phys. Soc. (London) B70 (1957) 669.
3. W.E. Spear, Proc. Phys. Soc. (London) B76 (1969) 826.
4. F.K. Dolezalek, in: J. Mort, D.M. Pai (Eds.), Photoconductivity and Related Phenomena, Elsevier, New York, 1976 (Chapter 2).
5. J.M. Marshall, Rep. Prog. Phys. 46 (1983) 1235.
6. J. Mort, I. Grammatica, M. Morgan, Appl. Phys. Lett. 40 (1982) 980.
7. S.O. Kasap, C. Juhasz, Solid State Commun. 63 (1987) 553.
8. S.O. Kasap, R.P. Thakur, D. Dodds, J. Phys. E: Sci. Instrum. 21 (1988) 1195.
9. S.O. Kasap, B. Polischuk, D. Dodds, Rev. Sci. Instrum. 61 (1990) 2080.
10. N. Theophanous, et al., J. Phys. E: Sci. Instrum. 21 (1988) 667.
11. B. Polischuk, S.O. Kasap, Meas. Sci. Technol. 2 (1991) 75.
12. I.P. Batra, K.K. Kanazawa, H. Seki, J. Appl. Phys. 41 (1979) 3416.
13. S.B. Berger, R.C. Enck, N.E. Scharfe, B.E. Springett, in: E. Gerlachand, P. Gross (Eds.), The Physics of Selenium and Tellurium, Springer, New York, 1979, p. 256.
14. J. Mort, D.M. Pai, in: J. Mort, D.M. Pai (Eds.), Photoconductivity and Related Phenomena, Elsevier, New York, 1976.
15. S.M. Vaezi-Nejad, Int. J. Electron. 62 (1987) 361.
16. F.W. Schmidlin, Phys. Rev. B 16 (1977) 2362.
17. L. Cheung, G.M. Foley, P. Fournia, B.E. Springett, Photogr. Sci. Eng. 26 (1982) 245.
18. C. Juhasz, M. Vaezi-Nejad, S.O. Kasap, J. Mater. Sci. 22 (1987) 2569.
19. S. Imamura, Y. Kanemitsu, M. Saito, H. Sugimoto, J. Non-Cryst. Solids 114 (1989) 121.
20. S.O. KasapA.S. Diamond (Ed.), Handbook of Imaging Materials, Marcel Dekker, Inc., New York/Hong Kong, 1991, p. 329.
21. B.E. Springett, in: Proc. 4th International Symposium. Uses of Selenium and Tellurium, Selenium–Tellurium Development Association, Inc, New York, 1989, p. 126.
22. W.E. Spear, J. Non-Cryst. Solids 1 (1969) 197.
23. J.M. Marshall, A.E. Owen, Phys. Status Solidi (a) 12 (1972) 181.
24. A.E. Owen, W.E. Spear, Phys. Chem. Glasses 17 (1976) 174.
25. R.C. Enck, G. Pfister, in: J. Mort, D.M. Pai (Eds.), Photoconductivity and Related Phenomena, Elsevier, New York, 1976.
26. M. Abkowitz, R.C. Enck, Phys. Rev. B 25 (1982) 2567.
27. G. Juska, J. Non-Cryst. Solids 137/138 (1991) 401.
28. G. Juska, K. Arlauskas, E. Montrimas, J. Non-Cryst. Solids 97/98 (1987) 559.
29. V.I. Mikla, Ph.D. Thesis, Odessa State University, 1984.
30. M.D. Tabak, P.J. Warter, Phys. Rev. 173 (1968) 899.
31. D.J. Gibbons, A.C. Papadakis, J. Phys. Chem. Solids 29 (1968) 115.
32. S.M. Vaezi-Nejad, Int. J. Electron. 62 (1987) 361.
33. H. Scher, E.W. Montroll, Phys. Rev. B 12 (1975) 2455.
34. K. Tanaka, Rev. Solid State Sci. 2/3 (1990) 644.
35. S.R. Elliott, J. Non-Cryst. Solids 81 (1986) 71.
36. K. Tanaka, MRS Int. Mtg. Adv. Mater. 12 (1989) 225.
37. V.I. Mikla, D.G. Semak, A.A. Kikineshi, Ukr. Phys. J. 25 (1980) 2021.
38. L.P. Kasakova, L. Toth, M.A. Taguyrdzhanov, Phys. Status Solidi (a) 71 (1982) K 107.
39. M. Itoh, K. Tanaka, Jpn. J. Appl. Phys. 34 (1995) 2216.
40. G. Pfister, H. Scher, Adv. Phys. 27 (1978) 747.
41. J.M. Marshall, Rep. Progr. Phys. 46 (1983) 1235.
42. V.I. Mikla, D.G. Semak, A.V. Mateleshko, A.A. Baganich, Phys. Status Solidi (a) 117 (1990) 242.
43. J. Schotmiller, M. Tabak, G. Lucovsky, A. Ward, J. Non-Cryst. Solids 4 (1970) 80.

44. V.I. Mikla, et al., J. Mater. Sci. 35 (2000) 4907.
45. V.I. Mikla, et al., Mater. Sci. Eng. B 64 (1994) 1.
46. V.I. Mikla, et al., J. Non-Cryst. Solids 246 (1994) 46.
47. V.I. Mikla, Phys. Status Solidi (b) 182 (1994) 325.
48. V.I. Mikla, et al., J. Phys. Condens. Matter 6 (1994) 8269.
49. V.I. Mikla, Phys. Status Solidi (a) 165 (1998) 427.
50. V.I. Mikla, et al., Phys. Status Solidi (a) 117 (1990) 241.
51. V.I. Mikla, V.V. Mikla, J. Non-Oxide Glasses 1 (2009) 1.
52. V.I. Mikla, V.V. Mikla, J. Non-Oxide Glasses 1 (2009) 24.
53. V.I. Mikla, V.V. Mikla, J. Mater. Sci. Mater. Electron. 20 (2009) 1059.

5 Xerographic Spectroscopy of States in Mobility Gap

Contents

Mobility measurements by the TOF methods considered in Chapters 3 and 4 are particularly important, but they cannot give information about the whole spectrum of states in the mobility gap of amorphous chalcogenides. Therefore, in addition to TOF, XTOF, IFTOF, TSC, and TSDC, other complimentary techniques that probe the gap states are needed. Xerographic techniques that were initially developed to characterize properties of electrophotographic (xerographic) receptors [1] seemed to be informative, suitable, and widely applicable for the study of amorphous thin films and photoconductive insulator thin films [2].

For further consideration of such a powerful instrument, it is helpful to provide information about the xerographic process (although the latter is well described in the scientific literature).

Xerography (or electrophotography) is a dry photocopying technique developed by Chester Carlson in 1938, for which he was awarded U.S. Patent 2,297,691 on October 6, 1942. Carlson originally called his invention "electrophotography." It was later renamed "xerography"—from the Greek roots ξηρός xeros "dry" and—γραφία—graphia "writing"—to emphasize that, unlike reproduction techniques then in use such as cyanotype, this process used no liquid chemicals.

Although Georg Christoph Lichtenberg invented a dry electrostatic printing process in 1778, Carlson's innovation combined electrostatic printing with photography. Carlson's original process was cumbersome, requiring several manual processing steps with flat plates. It was almost 18 years before a fully automated process was developed—the key breakthrough was use of a cylindrical drum coated with selenium instead of a flat plate. This resulted in the first commercial automatic copier,

Trap Level Spectroscopy in Amorphous Semiconductors. DOI: 10.1016/B978-0-12-384715-7.00005-X

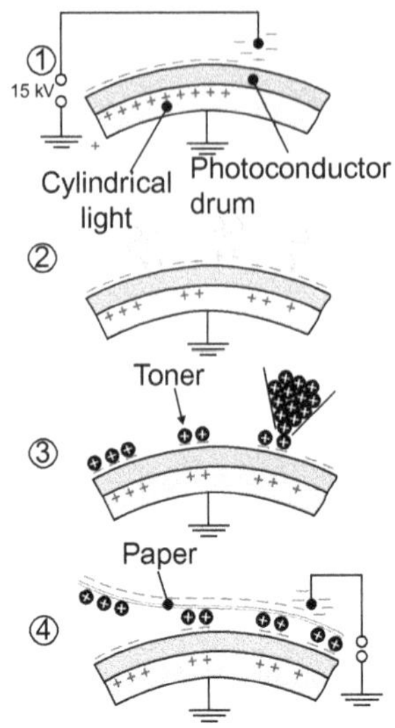

Figure 5.1 Illustration of the most important sequential steps used in xerography.

released by Haloid/Xerox in 1960. Today xerography is used in most photocopying machines and in laser and LED (light-emitting diode) printers.

Let us consider briefly the xerographic process (Fig. 5.1).

5.1 Schematic Overview of the Xerographic Photocopying Process

The first commercial use of xerographic photocopying was hand processing of a flat photosensor with a copy camera and a separate processing unit to produce offset lithographic plates. Today this technology is used in photocopy machines, laser printers, and digital presses such as the Xerox iGen3 and Xeikon presses, which are slowly replacing many traditional offset presses in the printing industry for shorter runs.

By using a cylinder to carry the photosensor, automatic processing was enabled. In 1960 the automatic photocopier was created and many millions have since been manufactured. The same process is used in microform printers and computer output laser or LED printers.

The steps of the process are described here as applied on a cylinder, as a photocopier. Some variants are considered within the chapter. Every step of the process has its own design.

A metal cylinder is mounted to rotate about a horizontal axis. This is called the drum. The end-to-end dimension is the width of print to be produced plus a generous tolerance. The drum in the copiers originally developed by Xerox Corporation was manufactured with a surface coating of amorphous selenium (more recently, ceramic or organic photoconductor (OPC)), applied by vacuum deposition. Amorphous selenium will hold an electrostatic charge in darkness and will conduct away such a charge under light. In the 1970s, IBM Corporation sought to avoid Xerox's patents for selenium drums by developing OPCs as an alternative to the selenium drum. OPCs are now preferred because they can be deposited on a flexible oval or triangular belt instead of a round drum.

Laser printer photo drums are made with a doped silicon diode sandwich structure with a hydrogen-doped, silicon light-chargeable layer; a boron nitride rectifying (diode-causing) layer that minimizes current leakage; and a surface layer of silicon doped with oxygen or nitrogen (silicon nitride is a scuff-resistant material).

The drum rotates at the speed of paper output. One revolution passes the drum surface through the following steps:

Step 1: **Charging.** An electrostatic charge is uniformly distributed over the surface of the drum in the dark by a corona-producing device correctly called a screened corotron or scorotron for short. How well the drum is charged depends on several variables. The most important are listed here:

1. The magnitude of the corona-produced current
2. The amount of this current reaching the drum surface
3. The uniformity with which the charge is deposited on the surface of the drum.

A well-designed corona device should charge the drum to any preselected voltage, evenly distribute the charge over the floating surface, and do so consistently for a long period of time. In such a way, the corona device is used to form the electrostatic images in the xerographic process. It is necessary to note that the corona device mentioned earlier is also widely used in xerographic spectroscopy. The authors provide some details about exploiting the corona device for the purpose of band-gap xerographic spectroscopy later.

Regarding the xerographic process, the output of a corona unit is limited by a control grid or screen. A charging effect can also be achieved with the use of a contact roller with a charge applied to it. The polarity is chosen to suit the positive or negative process. The positive process is used for producing black on white analog copies. The negative process is used for producing black on white from negative originals (mainly microfilm) and for all digital printing and copying. This is to economize on the use of laser light by the black writing or write-to-black exposure method.

Step 2: **Exposure.** The document or microform to be copied is illuminated and either passed over a lens or scanned by a moving light and lens, such that its image is projected onto and synchronized with the moving drum surface. Alternatively, the image may be flash exposed, using a xenon strobe, onto the surface of the moving drum or belt, fast enough to render a perfect latent image. Where there is text or an image on the document, the corresponding area of the drum will remain unlit. Where there is no image, the drum will be illuminated and the charge will be dissipated. The charge that remains on the drum after this exposure is a "latent" image and is a positive of the original document.

In a laser or LED printer, modulated light is projected onto the drum surface to create a latent image. The modulated light is used only to create the positive image, hence the term "black writing."

Step 3: **Development.** The drum is presented with a slowly turbulent mixture of toner particles and larger, metallic carrier particles. The carrier particles have a coating that during agitation, generates a triboelectric charge (just one form of static electricity), which attracts a coating of toner particles. The mix is manipulated with a magnetic roller to present a brush of toner to the surface of the drum/belt. By contact with the carrier, each neutral toner particle has an electric charge of polarity opposite to the charge of the latent image on the drum. The charge attracts toner to form a visible image on the drum. To control the amount of toner transferred, a bias voltage is applied to the developer roller to counteract the attraction between toner and latent image.

Where a negative image is required, e.g., when printing from a microform negative, then the toner has the same polarity as the corona in Step 1. Electrostatic lines of force drive the toner particles away from the latent image toward the uncharged area, which is the area exposed from the negative.

Early color copiers and printers used multiple copy cycles for each page output, using colored filters and toners. Modern units use only a single scan to four separate, miniature process units, operating simultaneously, each with its own coronas, drum, and developer unit.

Step 4: **Transfer.** Paper is passed between the drum and the transfer corona, which has a polarity that is the opposite of the charge on the toner. The toner image is transferred by a combination of pressure and electrostatic attraction from the drum to the paper. On many color and high-speed machines, it is common to replace the transfer corona with one or more charged bias transfer rollers (BTRs), which apply greater pressure and a higher quality image.

Step 5: **Separation or detack.** Electric charges on the paper are partially neutralized by AC from a second corona, usually constructed in tandem with the transfer corona and immediately after it. As a result, the paper, complete with most (but not all) of the toner image, is separated from the drum or belt surface.

Step 6: **Fixing or fusing.** The toner image is permanently fixed to the paper using either a heat-and-pressure mechanism (hot-roll fuser) or a radiant fusing technology (oven fuser) to melt and bond the toner particles into the medium (usually paper) being printed on.

Step 7: **Cleaning.** The drum, having already been partially discharged during detack, is further discharged by light, and any remaining toner that did not transfer in Step 6 is removed from the drum surface by a rotating brush, under suction, or a squeegee, known as the cleaning blade. In most cases, this "waste" toner is routed into a waste toner compartment for later disposal; however, in some systems, it is routed back into the developer unit for reuse. This process, known as toner reclaim, is much more economical but can possibly lead to a reduced overall toner efficiency through a process known as "toner polluting," whereby concentration levels of toner/developer having poor electrostatic properties are permitted to build up in the developer unit, reducing the overall efficiency of the toner in the system.

It is necessary to note that some systems have entirely abandoned the use of a separate developer (carrier). These systems, known as monocomponent, operate as described previously, but they use either a magnetic toner or fusible developer (however you wish to view it). This results in the complete removal of the need to replace wornout developer because the user effectively replaces it along with the toner. An alternative system, currently being developed by KIP from an abandoned line of research by Xerox, completely replaces magnetic toner manipulation and the cleaning system with a series of computer-controlled, and varying, biases. The toner is printed directly onto the drum, by direct contact with a rubber developing roller, which, by reversing the bias, removes all the unwanted toner and returns it to the developer unit for reuse.

The development of xerography has led to new technologies that some predict will eventually eradicate traditional offset printing machines. These new machines, which print in full CMYK color (printing terminology for cyan, magenta, yellow, and black), such as Xeikon, use xerography but provide nearly the quality of traditional ink prints.

5.2 Xerography in Animation

Ub Iwerks adapted xerography to eliminate the hand-inking stage in the animation process by transferring the animator's drawings directly to the cells. The first animated feature film to use this process was *One Hundred and One Dalmatians* (1961). At first, only black lines were possible, but in the 1980s, colored lines were introduced and used in animated features such as *The Secret of NIMH*.

In short, the high field (10^5–10^6 V/cm) due to corona charging is applied to sample films as the first step. Although electrostatic charging can be achieved in several ways (corona discharge, radioactive ionization of air, charge induction, charge transfer, etc.), the most practical and successful is undoubtedly corona discharge.

To investigate charge transport in high-resistivity materials, XTOF and xerographic discharge techniques have been used. The XTOF technique was previously described in Chapter 4. Here, we focus our attention on the xerographic discharge technique. Before considering this spectroscopic technique, however, the construction and performance of different types of corona device is described.

When a high voltage of several kilovolts is applied to a corona emitter, the field near the emitter exceeds the threshold field for air breakdown. Under this condition, the molecules near the emitter become ionized. The ions may carry a positive or negative charge, depending on whether the emitter is at a positive or negative voltage. If the insolating material (supported by earthed metal) is placed close to the corona emitter, some of the generated ions migrate to the floating surface of the sample and increase or decrease the surface charge density.

To improve the charging performance, the grounded electrodes are placed around the corona emitter. The structure is called the corona housing and helps direct the ion current and holds the corona emitter. Screen grids are then inserted between the corona emitter and the sample to provide better control of charging; this is the corona grid. Several variables determine the efficiency of sample charging. We note here only the most important:

a. the uniformity of the charge deposited on the sample surface;
b. the amount of corona current that reaches the sample surface.

Reasonably, the assessment of electrostatic characteristics of the sample examined depends on its particular applications. As for electrophotography, the following may be listed: measuring the charge acceptance (or the maximum surface charging potential), dark decay of the surface potential of charged (in darkness) sample, light decay, and the so-called residual potential (in other words, the potential that remains after photodischarge of the sample).

There are two more or less exploited types of corona devices—a pin corona discharge device and a wire corona discharge device. These devices are well described by Vaezi-Nejad and Juhasz [3].

A pin corona discharge device consists of a stainless steel rod with a sharp conical tip as the corona emitter, a brass hollow cylinder as the corona case, and a perspex cup as the corona holder. Experimental results show that the pin corona current distribution is not uniform, which is one disadvantage of a pin corona device.

The wire corona discharge device consists of a hollow stainless steel cylinder (inner diameter = 3–5 cm, length = 10.7 cm, and thickness = 0.16 cm) as the corona case with a 5.7 cm × 3.0 cm window. Two perspex slabs of dimensions 5.5 cm × 3.4 cm × 1.25 cm are used to hold the corona wire. The primary requirement for a corona wire emitter is that it should produce ions by means of a corona discharge at a reasonable voltage, typically 3–7 kV. In order to achieve a high electric field, the wire diameter must be small (less than 100 μm). Tungsten is particularly suitable as the material for corona emitter because of its resistance to the harsh environment created by the corona discharge, such as ultraviolet light and various nitrogen–oxygen compounds produced in the discharge. Another advantage of this material is that it is mechanically strong so as to resist breakage caused by the stress of stringing, handling, and cleaning. The corona onset voltage of 1 kV is easily achievable. It is possible to improve single-corona (described earlier) charging characteristics by adding an identical tungsten wire to the device, thus forming a double-wire corona discharge. The wires were 3.7 cm long and spaced 0.5 cm apart [3].

In an attempt to find most efficient corona discharge device for xerographic–spectroscopic purposes, it seems necessary (as the author [4] did) to try various device configurations utilizing sharp pin(s) and wire(s) as the corona emitter. The schematic sketches of three corona devices are shown in Fig. 5.2.

Experimental results obtained in Ref. [4] illustrate clearly that devices based on thin wires provide a more uniform charge distribution and thus may be recommended for XTOF spectroscopy. In order to control the sample initial voltage, a biased grid is inserted between the corona emitter and the floating surface of the sample.

After the sample is charged, it is necessary to exactly measure the surface potential at various stages. Reasonably, these are the following:

1. The initial voltage of the sample for calculation of drift mobility.
2. The sample voltage after photoexcitation. One can estimate the total charge injected into the sample, subtracting the sample voltage after photoexcitation (photodischarge) from the initial voltage.
3. The final value of the sample voltage measured after photoexcitation. This is known in the literature as the residual voltage and is due to deeply trapped photoinjected carriers when they transit across the sample. If the sample is repeatedly charged–discharged, the residual voltage builds up with the number of cycles and saturates. In this case, one can use the saturated residual voltage and extract information about the deep traps.

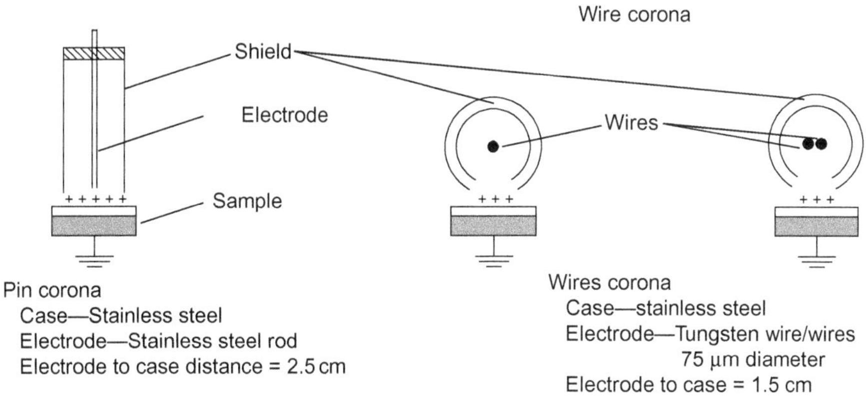

Figure 5.2 Schematic sketches of different corona devices [4].

After charging, the decay of the open-circuit surface potential is measured. From these measurements, important information can be extracted. In the past few decades, the xerographic probe technique has become a very popular and unique means to characterize electronic gap states. In particular, a map of states near mid-gap is determined by a time-resolved analysis of the xerographic surface potential.

5.3 Xerographic Dark Decay and Photoinduced Effects

Analysis of the time and temperature dependent decay of the surface voltage on an amorphous film after charging, but prior to exposure (xerographic dark decay), and of residual decay after exposure can (in combination) be used to map the density of states.

An optimal photoreceptor design will require, among other factors, high charge acceptance, slow dark discharge, low first and cycled-up (saturated) residual voltages, and long carrier ranges ($\mu\tau$). The latter factor has been addressed earlier. There are essentially three important types of xerographic behavior—generally termed the dark discharge, first-cycle residual, and the cycled-up residual voltage—that must be considered in evaluating the electrophotographic properties of a-Se and its alloys. It should be stressed that these three parameters are extremely informative in the sense that they map band-gap states.

In xerographic measurements, as illustrated in Fig. 5.3, the sample is corona-charged to a voltage V_0 and then exposed to a short wavelength (absorption depth $\delta \ll L$) step illumination. At the end of the illumination, there is a measurable surface potential, termed the residual potential V_r, because of the bulk trapped charges. If positive charging is used, then V_r is due to trapped holes in the bulk of the specimen.

Dark discharge rate must be sufficiently low to maintain an ample amount of charge on the photoreceptor during the exposure and development steps. A high dark decay rate will limit the available contrast potential. The residual potential remaining after the xerographic cycle must be small enough that it does not impair the quality

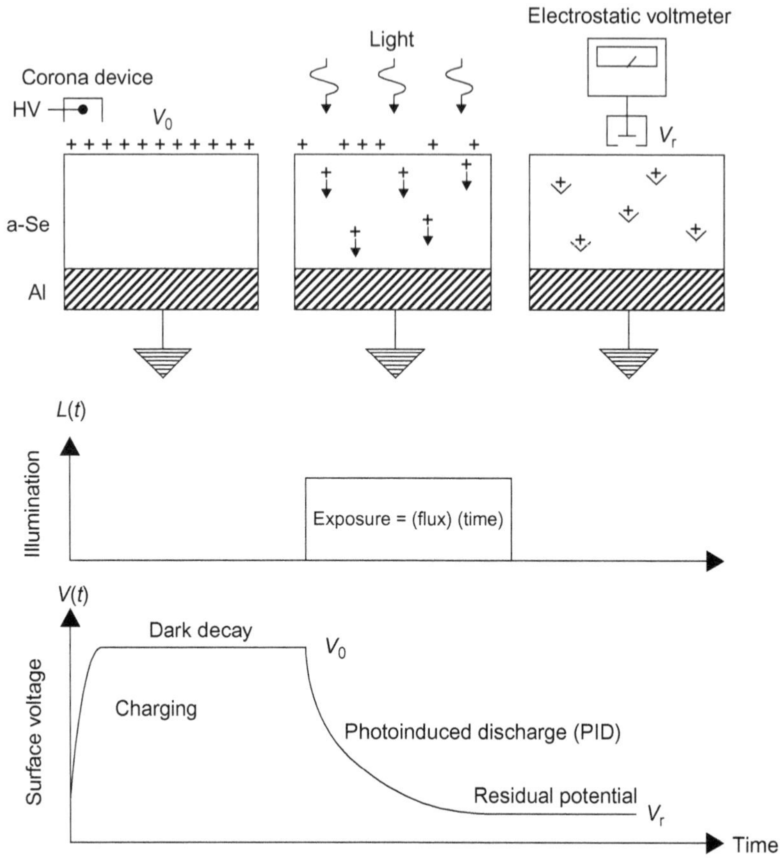

Figure 5.3 The xerographic measurements technique (schematic illustration) [2].

of the electrostatic image in the next cycle. Over many cycles, the cycled-up residual potential should be small to avoid deterioration in the copy quality after many cycles.

In the case of a-Se, these xerographic properties have been extensively studied. In addition to the magnitude of the saturated residual voltage, the rate of decay and the temperature dependence of the cycled-up residual potential are important considerations, because they determine the time required for the photoreceptor to regain its first-cycle xerographic properties.

Figure 5.4 displays the simplest experimental setup for xerographic measurements. The rotating photoreceptor drum is charged at station A by a corotron device. The surface potential is measured at B, and the photoreceptor is then exposed to a controlled wavelength and intensity illumination at station C, following which its surface potential is measured again at a station D. In some systems, the surface potential is also monitored during exposure at C via a transparent electrometer probe to study the photoinduced discharge characteristics. Normally, the charging voltage,

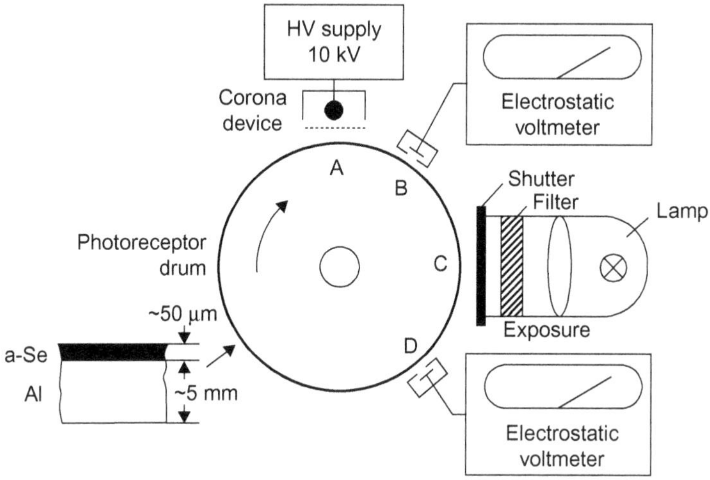

Figure 5.4 Simplified schematic diagram of a xerographic measurement. The photoreceptor is charged at A and exposed at C. Its surface potential is measured before and after exposure at B and D [2].

speed of rotation, and exposure parameters such as energy and wavelength are user-adjustable (Fig. 5.5).

Figure 5.6 shows typical positive and negative dark discharge curves for pure a-Se films prepared under different conditions [2], where it can be seen that the dark discharge rate depends on substrate temperature.

The currently accepted model for the dark decay in a-Se films involves substrate injection as well as bulk thermal generation and depletion. The latter process involves thermal generation of holes in the bulk and their sweep out from the sample by the internal field, leaving behind a bulk negative space charge. In other words, the decay of surface potential on pure and alloyed amorphous selenium films has been found to be controlled by the depletion discharge process [5–11]. In essence, the xerographic depletion discharge model is based on bulk thermal generation involving the ionization of a deep mobility gap center to produce a mobile charge carrier of the same sign as the surface charge and an oppositely charged ionic center. Assuming positive charging (as in earlier discussions), a mobile hole would be thermally generated and the ionized center would be negative. As thermally generated holes are swept out by the electric field, a negative bulk space charge builds up with time in the specimen, causing the surface potential to decay with time. If the buildup in the bulk negative charge density is spatially uniform, the internal electric field falls linearly with distance from the top surface. At a certain time, called the depletion time t_d, the electric field F at the grounded end of the sample becomes zero. From that time onward, the field will be zero at a distance $X(t) < L$, the sample thickness; consequently, there will be a neutral region from X to L inasmuch as holes generated in $0 < x < X$ and arriving into $X < x < L$ will not be swept out. The shrinkage of the depleted volume

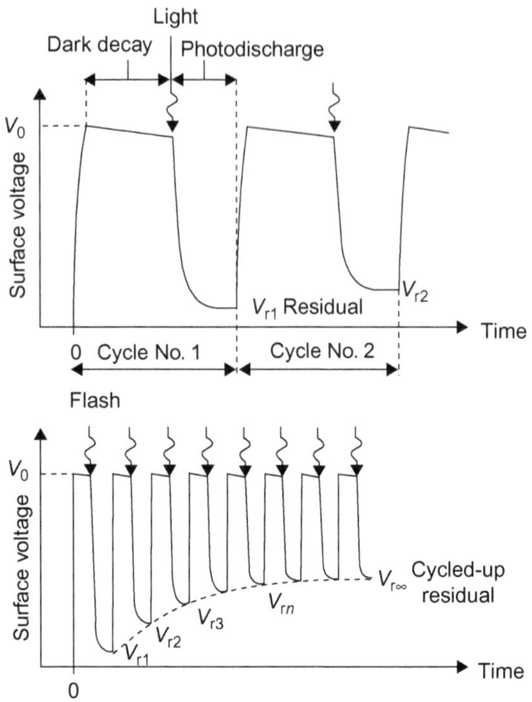

Figure 5.5 Typical photoreceptor behavior through xerographic cycles showing dark decay, first-cycle residual potential V_{r1}, and cycled-up residual potential V_{rn} after many cycles [2].

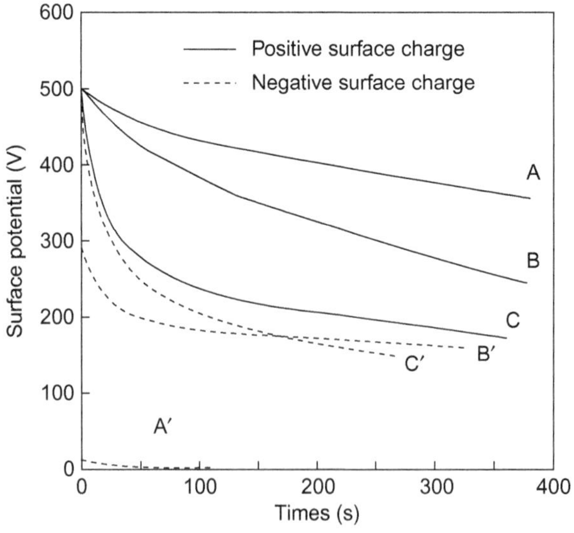

Figure 5.6 Dark discharge of surface potential on a-Se layers. A, B, and C involve a-Se deposited under different substrate temperature (T_s) conditions: A and A′ at T_s = 75°C; B and B′ at T_s = 50–60°C; C and C′ at T_s = 25–50°C and uncontrolled [2].

with time $t > t_d$ means that t_d marks a functional change in the dark decay rate and, therefore, is readily obtainable from dark discharge experiments (Fig. 5.7).

Under low charging voltages, the depletion time indicates the time required for the surface potential to decay to half its original value. Under high charging voltages, however, field-enhanced emission from the deep mobility gap centers also plays an important role, and the surface potential initially decays at a much faster rate so that at the depletion time of the surface potential is, in fact, less than half the initial value. Figure 5.8 shows the dependence of the depletion time t_d and the half-time $t_{1/2}$ on the charging voltage V_0, where it can be seen that at the highest charging voltages, there is no improvement in $t_{1/2}$, with a further increase in the charging voltage V_0. Inasmuch as the dark decay in Se-based alloys is a bulk process, the rate of discharge increases with the square of thickness ($dV/dt \sim L^2$) and can be reduced only by using thin layers. The latter concept naturally leads to the design of multilayer photoreceptor structures.

The origin of the deep localized states in the mobility gap that control the dark decay has been attributed to structural native thermodynamic defects [12]. Thermal cycling experiments show that the response of the depletion time to temperature steps is retarded, as would be expected when the structure relaxes toward its metastable liquid-like equilibrium state. As the structure relaxes toward the equilibrium state, t_d decreases further until the structure has reached equilibrium. The only possible inference is that t_d must be controlled by structure-related thermodynamic defects. The generation of such defects is, therefore, thermally activated. We should note that because the depletion discharge mechanism involves the thermal emission of carriers

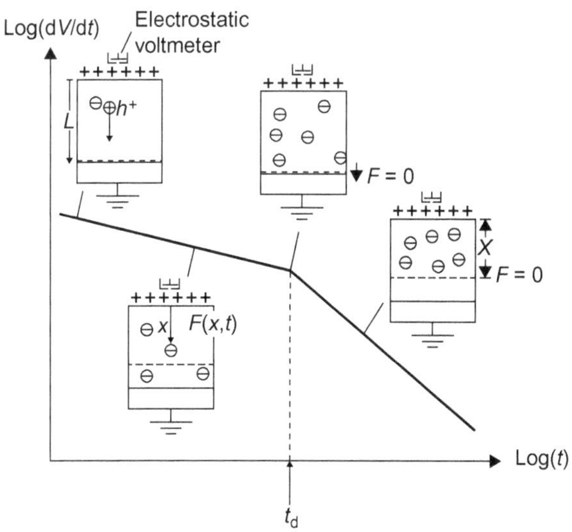

Figure 5.7 Typical log–log plot of the dark discharge rate versus time for an amorphous Se–Te film [13].

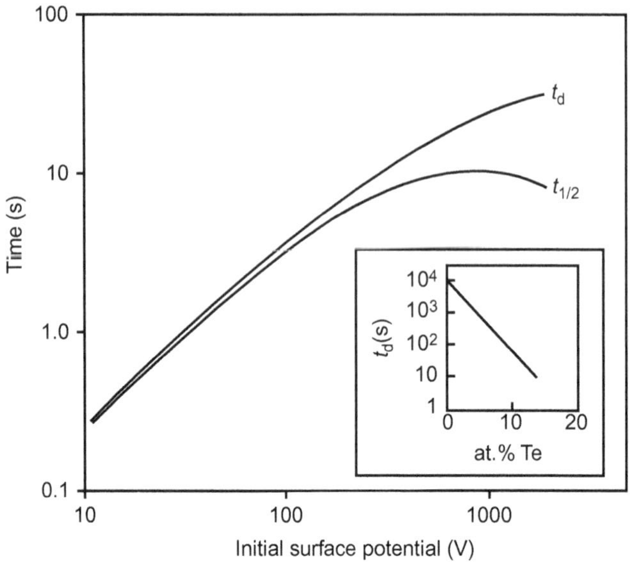

Figure 5.8 Log–log plots of the depletion time t_d and time for the surface potential to decay to its half-value $t_{1/2}$ versus charging voltage V_0 for an amorphous Se:Te 13 wt% photoreceptor film of thickness 70 μm (from Ref. [8]). The inset shows the dependence of the depletion time t_d on the Te content (from Ref. [9]).

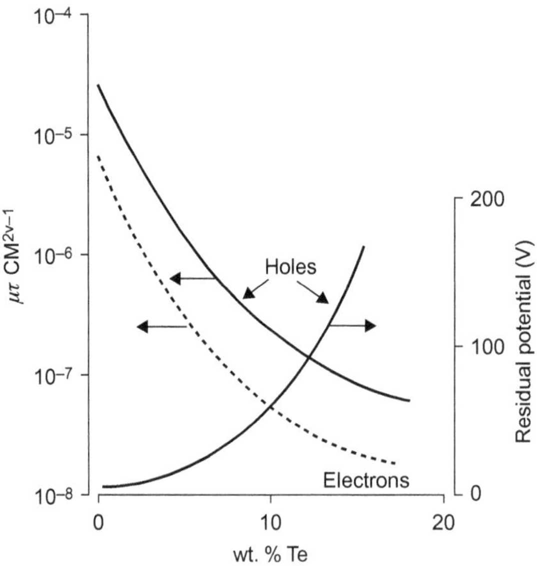

Figure 5.9 Hole and electron drift mobility lifetime product $\mu\tau$ and residual potential versus Te content in a-$Se_{1-x}Te_x$ films. The $\mu\tau$ product was xerographically measured by Abkowitz and Markovics [14].

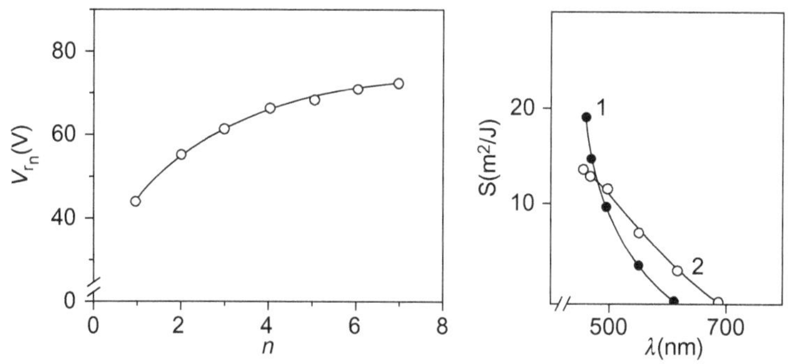

Figure 5.10 Residual potential V_{rn} as a function of xerographic cycle n for a-Se:Sb films ($L = 50\,\mu m$).

from deep localized states, it is strongly temperature dependent. For example, t_d increases in an approximate Arrhenius fashion with decreasing temperature [6]. In addition to the deterioration of the dark decay, there is an increase in the residual potential for a-$Se_{1-x}Te_x$ alloys.

Figure 5.9 displays the $\mu\tau$ product for holes and electrons [13].

This parameter was determined from xerographic residual potential in a-$Se_{1-x}Te_x$ monolayer films. As one can see, even with very little Te alloying, there is a considerable rise in both hole and electron deep traps. The relationship between the trapping time and the residual potential has been evaluated by several authors (see, for example, Refs. [12, 15]). It can be seen that once the Te concentration exceeds 12 wt% Te, the residual potential is more than an order of magnitude larger than typical values for pure selenium.

Returning to the depletion discharge, it should be stressed that with thick films and a good blocking contact between a-Se and the preoxidized aluminum substrate, the depletion discharge process dominates. There are several reasons that amorphous selenium possesses good dark decay characteristics:

1. There are not many deep localized states in the mobility gap of pure amorphous selenium.
2. The energy location of these localized states is deep in the mobility gap. Therefore, the thermal generation process of holes (or electrons) from these centers is slow.
3. Injection from the substrate can be reduced substantially by using oxidized Al substrate.

Figure 5.10 displays the buildup of the residual voltage V_{r5} on an a-Se:Sb film with number of xerographic cycles.

If blue light is used for the discharge process, then the absorption is very close to the charged surface, and one can assume that the discharge process involves the transport of photogenerated holes through the bulk. Trapping of these holes in the bulk then results in the observed first residual potential V_{r1}. In the case of amorphous selenium photoreceptor films, it has been found that V_{r1} is predicted by the Warter expression [16]

$$V_{r1} = \frac{L^2}{(2\mu_d\tau_0)}$$

where L is the film thickness, μ_d is the hole drift mobility, τ_0 is the hole lifetime, and $\mu_d\tau_0$ is the hole range.

It can be seen from Fig. 5.11 that as the xerographic cycle is repeated many times at a constant repetition frequency, the residual voltage rises and eventually saturates. The saturated residual voltage V_{rs} is much larger than the first-cycle residual voltage V_{r1}. Both the first residual and the cycled-up saturated residual potential, V_{r1} and V_{rs}, are sensitive to preillumination as well as to temperature and alloying. For example, when a-Se films are preilluminated with white light, the buildup of the residual potential occurs more rapidly toward a much higher saturated residual potential. Furthermore, the parameters mentioned (residual potentials V_{r1} and V_{rs}) increase with exposure time. This experimental fact is well established by the authors. Clearly, exposure to white light generates an appreciable concentration of deep hole traps. There are two possible explanations for the saturation of the residual voltage, discussed next.

The observed saturation may be due to the dynamic balance between trapping and release of charge carriers as the xerographic cycle is repeated. As an alternative variant, it may be due to the filling of the deep trap population so that the saturated residual potential is given by

$$V_{rs} = \frac{L^2 e N_t}{(2\varepsilon_0\varepsilon)}$$

where N_t is the concentration of deep traps, and ε_0 and ε are, respectively, are the absolute permittivity and relative permittivity of the photoreceptor material.

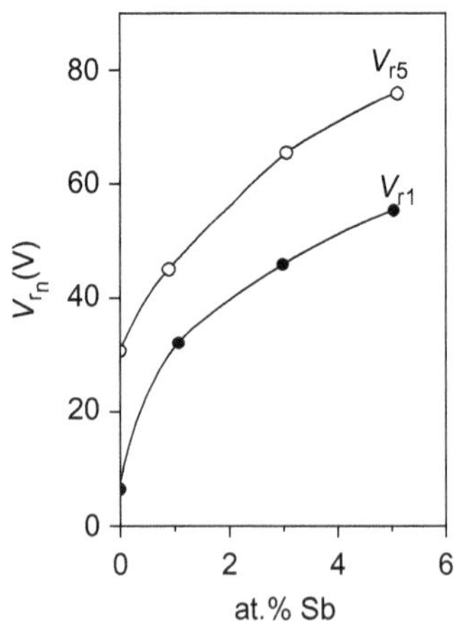

Figure 5.11 Dependence of the first and firth residual potentials (for pure a-Se see Ref. [17]).

The rate of decay and the temperature dependence of the saturated voltage can be used to obtain the concentration and energy distribution of the deep traps responsible for the residual potential. Thus, V_{rs} provides a useful means of studying the nature of deep traps in amorphous semiconductors and has been successfully used to derive the energy distribution of deep localized states in the mobility gap of both a-Se and a-Si:H [10, 18].

One can assume that the saturated residual potential, at the end of a large number of cycles, decays. As thermal release proceeds, holes are emitted and swept out from the specimen, resulting in a decrease in the measured surface potential. The decay rate of the saturated potential is strongly temperature dependent due to thermal release from deep mobility gap centers, located approximately 0.9 eV above E_v for holes. The discharge of the saturated potential due to electron trapping occurs much more slowly. The reason is that the energy depth of electron traps from E_c is about 1.2 eV, which is greater than that of hole traps from E_v.

References

1. R.M. Schaffert, Electrophotography. Society of Photographic Scientists and Engineers, Focal Press, London, 1975 (Chapters 1 and 2).
2. S.O. Kasap, in: A.S. Diamond (Ed.), Handbook of Imaging Materials, Marcel Dekker, Inc., New York/Hong Kong, 1991, p. 329.
3. S.M. Vaezi-Nejad, C. Juhasz, Int. J. Electron. 67 (1989) 437.
4. S.M. Vaezi-Nejad, Int. J. Electron. 62 (1987) 361.
5. M. Abkowitz, G.M.T. Foley, J.M. Markovics, A.C. Palumbo, Appl. Phys. Lett. 46 (1985) 393.
6. M. Baxendale, C. Juhasz, SPIE Proc. (1990) 1253.
7. S.O. Kasap, M. Baxendale, C. Juhasz, IEEE Trans. Indust. Appl. 27 (1991) 620.
8. S.O. Kasap, J. Electrostat. 22 (1989) 69.
9. M. Abkowitz, F. Jansen, A.R. Melnyk, Phil. Mag. B51 (1985) 405.
10. M. Abkowitz, J. Non-Cryst. Solids 66 (1984) 315.
11. M. Abkowitz, J.M. Markovits, Solid State Commun. 44 (1982) 1431.
12. S.O. Kasap, J. Phys. D25 (1992) 83.
13. M. Abkowitz, R.C. Enck, Phys. Rev. B25 (1982) 2567.
14. M. Abkowitz, J.M. Markovics, Solid State Commun. 44 (1982) 1431.
15. K.K. Kanazawa, I.P. Batra, J. Appl. Phys. 43 (1972) 1845.
16. P.J. Warter, Appl. Optic. Suppl. 3 (1969) 65.
17. M. Abkowitz, R.C. Enck, Phys. Rev. B27 (1983) 7402.
18. S.O. Kasap, V. Aiyah, B. Polischuk, A. Bhattacharyya, Z. Liang, Phys. Rev. B43 (1991) 6691.

6 Photoinduced Effects on States in the Mobility Gap

Contents

6.1 Introduction

In this chapter, the effect of preexcitation with the light of band-gap energy on trapping and thermal generation is examined in selenium and selenium-rich As–Se alloy films by several techniques. Results suggest that excess carrier trapping and dark-carrier generation are controlled by deep defect centers whose population can temporarily be altered by photoexcitation.

6.2 Steady-State Photocurrents

There is a strong evidence to suggest that most of the reported photoinduced optical and structural changes affect the band-gap states. Several research groups have devoted considerable efforts to experiments that probe the gap states near mobility edges and within mobility gap (see, for example, Ref. [1] and references therein). This is not a simple matter; no single experiment gives complete and unequivocal information, thus making it necessary to bring together data from a wide range of measurements, stationary as well as transient.

First, we consider the associated changes in the photoelectronic properties of the samples. The spectral characteristics of photoconductivity of the samples display a red shift after irradiation. Such behavior of the photoconductivity is not surprising, because it is in full agreement with the shift in the absorption edge. Additionally, the photoconductivity decreases after photodarkening [2]. The decrease may be attributed to the creation of new defect states or altering the existing localized states.

Trap Level Spectroscopy in Amorphous Semiconductors. DOI: 10.1016/B978-0-12-384715-7.00006-1

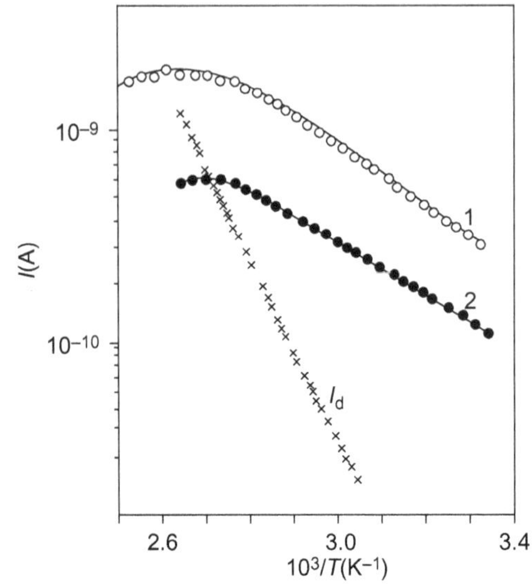

Figure 6.1 Steady-state photocurrent in an a-As_2Se_3 sample at 700 nm illumination with approximate intensity of 10^{14} photons/cm^2/s as a function of inverse temperature (curve 1). Also shown is the effect induced by laser irradiation ($\lambda = 633$ nm) at $T = 100$ K (curve 2).

Lux–ampere characteristics for annealed films are characterized by a sublinear dependence $I_{ph} \sim E^n$ in a wide light intensity range, n being approximately equal to 0.6 for As_2Se_3. After darkening, the index n increases.

The temperature dependence of the steady-state photocurrent is generally quite similar to that observed widely in chalcogenides [3]. Figure 6.1 shows that photoconductivity is an activated process and varies exponentially with $1/T$. The activation energy from the plot's $\log I_{ph}$ versus $10^3/T$ turns out to be 0.31 eV. Also plotted in the figure is the variation in dark conductivity with temperature, which is also an activated process. Because it is generally accepted [3] that in chalcogenide semiconductors, recombination will be mediated by the charge defect states, we may analyze the temperature dependence of the steady-state photocurrent in terms of energy scheme involving acceptor-like levels in the gap. We believe the low-temperature "bimolecular" slope ΔE_b represents half the distance between the valence band and acceptor-like level. The observed differences in ΔE_b with exposure ($\Delta E_b \approx 0.22$ eV in irradiated samples) therefore indicate that illumination influences the gap-state density in the lower half of the band-gap.

The preceding changes in the photocurrent of amorphous chalcogenides may be related to specific changes in electronic gap states, which act as trapping and recombination centers and, therefore, limit the photoconductivity.

6.3 Light-Induced Effects on Photocurrent Transients

The most unique and intriguing features of chalcogenide vitreous semiconductors are the photoinduced changes appearing as a nearly parallel shift of the optical absorption edge to lower energy (so-called photodarkening effect) after irradiation with band-gap light. Such irradiation also causes changes in a wide variety of other properties—mechanical, physicochemical, photoelectronic, and so forth [4]. The latter provides a valuable opportunity for the evaluation of light-induced changes of the gap states. Over the past few years, however, there has been very limited systematic study of photoinduced effects on states, especially deep-lying states in the band-gap. Most photodarkening studies have focused on binary and ternary chalcogenide alloys, and relatively little is known about the characteristics of elemental and chalcogen-rich glasses. On the other hand, elemental and chalcogen-rich amorphous semiconductors serve as useful model systems for studying the influence of photodarkening on physical properties. In this chapter, we consider photoinduced changes of deep levels in selenium layers with arsenic additions up to 20%.

Currently, two classes of transient photocurrent experiments can be distinguished. On the one hand, there is unipolar method. The most common is the TOF experiment developed by Spear [5]. In the TOF method, the carriers are created near one electrode of a sandwich configuration. This configuration makes it possible to distinguish electron and hole motion. The carrier mobility is determined by the position of the break in the current pulse. Therefore, precise knowledge of concentration of the transiting carriers is not necessary. To observe the transit pulse in the TOF experiment, the electrode must form a blocking contact; i.e., it must be enable to inject carriers. To prevent distortion of the field, the excess charge introduced should be less than CV, where C is the capacitance of the sample and V is the voltage across the sample.

On the other hand, there is the bipolar measurement of transient photocurrent (due to electrons and holes) performed in the gap-cell configuration (see, for example, Refs. [6, 7]). In these experiments the sample is illuminated uniformly in the direction transverse to the electrodes, creating an excess conductance G. Injecting contacts supply the excess current $I_{ph} = GV$ for a bias voltage V. There is no buildup of charge at the electrodes. As a result, signal averaging with repetitive pulses is easy. When analyzing coplanar transient photocurrents, the different drift mobility of electrons and holes, together with the recombination, must be taken into account.

Let us first consider the results of TOF measurements on Se-rich amorphous chalcogenides [8].

The sample films used in all TOF studies were typically 5–50 μm thick. These were prepared by vacuum evaporation of As_xSe_{1-x} alloys at a fixed rate of approximately 1 μm/min onto glass substrates. The substrates were coated with SnO_2 or Al blocking contacts. All samples were annealed at room temperature for approximately 100 h and stored in the dark at least 48 h before measurement. Where sandwich geometry was required (photocurrent transients), semitransparent gold contacts were slowly evaporated onto the free surface of the specimens. Band-gap illumination was provided by a mercury lamp and a He–Ne laser. The transient transport measurements were performed using conventional (small-signal) TOF technique. The drift of the sheet of carriers initially photoexcited at the Au contact by a pulsed (5 ns duration, 336.1 nm wavelength) nitrogen laser is time resolved in a uniform applied field. All measurements were performed in the current mode, in which case the time constant of the external circuit was kept short compared to the transit time to ensure the constant voltage across the sample during the carrier drift. When dispersion entirely masks the transit time cusp, we determine the statistical transit time from a double-logarithmic plot of the algebraically decaying current.

Two types of newly developed [1, 9] xerographic measurements are discussed in the following section: dark decay and cycled-up xerographic residual voltage. The dark-decay experiments were carried out by charging the surface of the sample from a corotron device and then measuring the surface potential under an electrostatic probe. In the cycled-up residual voltage experiments, the surface voltage following photodischarge in the xerographic cycle is monitored as a function of the number of cycles.

At room temperature, the amorphous Se (a-Se) films exhibit very slight but definite photodarkening. The effect consists of a decrease of transmission ($\approx 2\%$ in magnitude) after illumination, which was spontaneously restored as excitation was switched off. It was found that the restoration period increases with the arsenic concentration. In other words, the photoinduced changes are transient (dynamic) and not permanent.

In order to observe the effect of preillumination (exposure conditions were chosen identical to those that cause transmission changes) on microscopic parameters of states, measurements of the transport characteristics were performed. We first consider the carrier drift and space charge as it develops in alloy films As_xSe_{100-x} with $0 \leq x \leq 15$ at.%, in which the transport of holes and electrons is observable [3, 8]. Figure 6.2 shows a typical nondispersive trace of the transient electron current in dark-rested and preexposed pure selenium. The room-temperature value of the drift mobility $\mu^e \approx 6 \times 10^{-3}$ cm^2/Vs is in remarkably good agreement with reported values and seems to be unaffected by irradiation. We observe only a small decrease of the current and a slightly changed shape of the signal. There is a close correlation between the recovery of optical parameters and the current in exposed samples. For example, the transmission of irradiated a-Se approaches its virgin value after about 7 min. The same time is necessary for relaxation of the electron drift transient. Note that in case of the hole, photocurrent transients of about 2 min are sufficient for restoring the original signal.

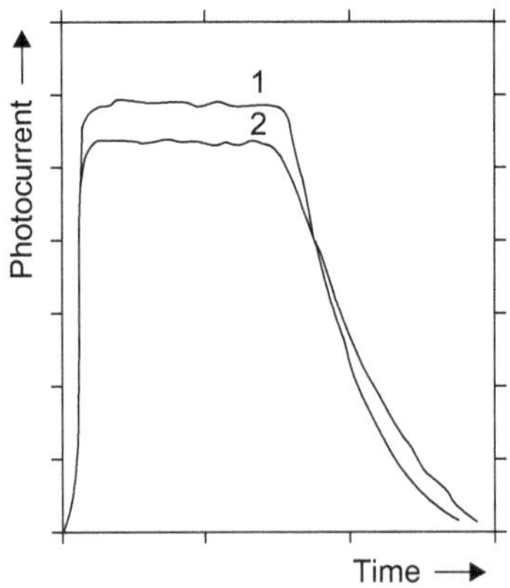

Figure 6.2 A typical oscilloscope trace of the TOF photocurrent electron signal in (1) dark-rested and (2) exposed a-Se. $d = 4.2\,\mu$m, $E = 14$ V/μm, $T = 293$ K, 0.5 μs/div.

As alloying results in decreasing both the electron and hole mobility μ^e, μ^h, and it substantially changes the shape of hole transients. At low arsenic concentrations (0–2 at.%), the hole transit pulses are well defined and retain the shape of pure Se. In the range 2 at.% $\leq x \leq$ 6 at.% As, the hole response is absent, and only a lifetime-limited signal is observed in accordance with the reported [8, 10] behavior. When raising the As concentration above 6 at.%, the hole signal reappears, but it consists of an initial fast response, a quasi-stationary region (plateau), and a long featureless tail typical for current transients observed in As_2Se_3. Band-gap irradiation of the sample leads essentially to a decrease in the current level, whereas the transit time t_T remains unchanged. If, after light exposure, the sample is allowed to rest in the dark, signal recovery takes place [2, 8] (Fig. 6.3).

Figure 6.4 shows the relaxation function $\psi(\tau)$ of irradiated As_xSe_{100-x} alloy films, determined as follows. The sample irradiated with light of photon energy $h\nu \geq E_{opt}$

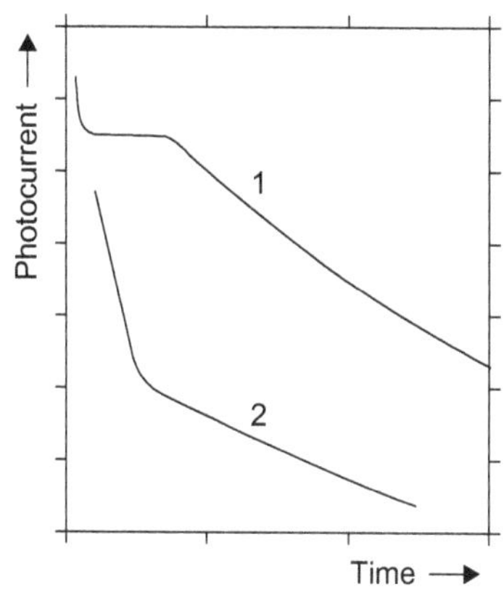

Figure 6.3 Oscilloscope trace of the TOF photocurrent hole signal in (1) dark-rested and (2) exposed a-$As_{0.10}Se_{0.90}$. d = 4.2 μm, E = 14 V/μm, T = 293 K, 0.5 μs/div.

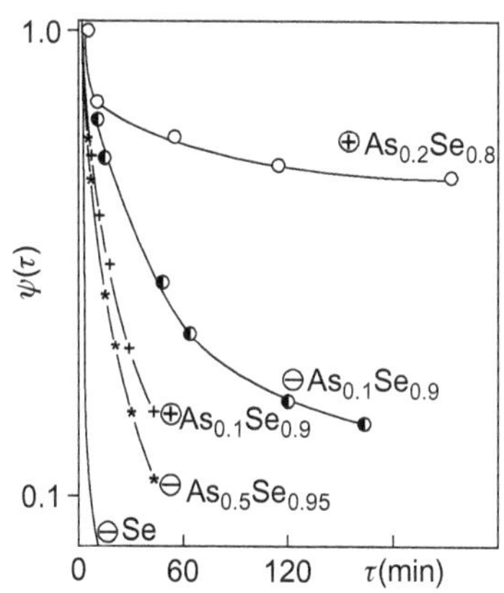

Figure 6.4 The relaxation function for hole (electron) metastable traps.

for a fixed time was allowed to rest in the dark for a time τ (experimental variable). The current signal $I \sim f(t)$ was recorded as a function of $\psi(\tau)$; $\psi(\tau)$ was determined as

$$\psi(\tau) = (I_r - I_\tau)/(I_r - I_p)$$

where I_r and I_p are the current values of dark-rested and exposed samples, respectively, immediately after illumination (current traces were chosen at $t = t_T$). Fully analogous results were obtained for the relaxation function $\psi(\tau)$ determined from charge collection experiments.

The increase of the As content and/or lowering of the temperature delays the relaxation both of electron and hole transit pulses in photosensitized samples. At the same time, asymmetry of the relaxation function for electrons and holes was clearly detected. From the data presented, it is obvious that the transit time remains unchanged by photodarkening for all investigated compositions. Consequently, shallow states that control charge transport and define the activation energy E_μ ($E_\mu^e \approx 0.33$ eV, $E_\mu^h \approx 0.25$ eV in a-Se, $E_\mu^h \approx 0.4$–0.6 eV in As_2Se_3) of the mobility should not undergo photoinduced changes. At the same time, the observed change of the current shape, the appearance of a long tail, and the decreasing signal height indicate a photoinduced change to deep ($E_t > E_\mu$) centers. Such a behavior is possibly related to enhanced carrier trapping by deep levels after photosensitization. Figure 6.5 supports this suggestion. It is apparent from the figure that preillumination leads not only to a decreasing magnitude of the photocurrent transient (in this case for electrons in an $As_{10}Se_{90}$ alloy film), but also to an abrupt transformation of the signal character from a pulse with well-defined transit time (curve 1, Fig. 6.5) to a featureless lifetime-limited pulse (curve 2, Fig. 6.5). The considerable decrease of transit pulse height was accompanied by a significant alteration of the current shape: the absolute value of the initial (pretransit) current slope

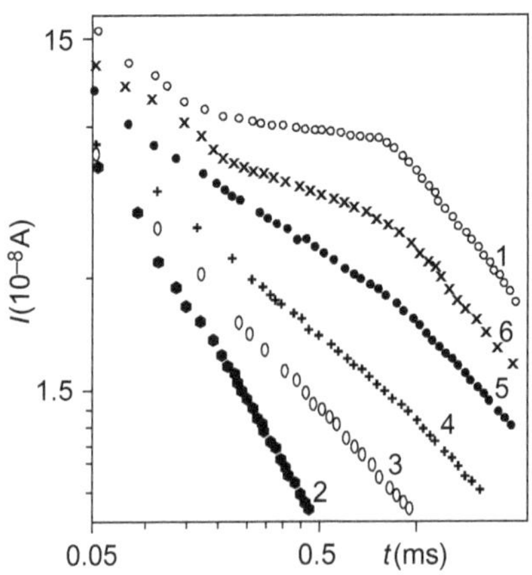

Figure 6.5 Transient electron current in $As_{0.10}Se_{0.90}$ samples ($d = 4.4\,\mu m$, $E = 3.7\,V/\mu m$, $T = 293$ K): data 1, dark rested; data 2, exposed; data 3, 4, 5, and 6, exposed and then dark rested for 5, 10, 50, and 160 min, respectively.

$I \sim t^{-(1-\alpha)}$ in double-logarithmic representation (for comparison, see corresponding curves, Fig. 6.5). If, after preillumination, the film is allowed to relax in the dark, the previous (dark-rested) shape of the transient current is fully recovered; i.e., the pulse height and shape return to values characteristic of the unexposed specimen. It is thus found that the charge-carrier lifetime with respect to deep trapping (τ_L) decreases approximately by an order of magnitude. For example, $\tau_L \leq 0.5$ ms for electrons in the previously illuminated sample $As_{0.10}Se_{0.90}$.

In an earlier study of transport in annealed As_2Se_3 films, we observed a similar behavior of photoinjected carrier transients under the conditions of photodarkening [2]. The main difference in the results presented here is that in As_2Se_3 films, optical and transport parameters once changed by irradiation never relax spontaneously to the equilibrium state. Restoration is possible only by annealing near the glass transition temperatures. Further investigations show that band-gap light leads to a pronounced variation of the space charge developing in the sample. Figure 6.6 shows typical transient hole photocurrents induced in As_2Se_3 films before and after photodarkening. When transit measurements are repeated, the current level diminishes progressively (current traces 1 and 2, 3 and 4, Fig. 6.6). It is obvious that after photodarkening, the space charge buildup is significantly intensified.

Drift mobility experiments may be affected by two kinds of space charge effect. The first type requires is the presence of deep centers in the gap [5]. These centers gradually accumulate charge with each drift pulse, especially near the surfaces of a specimen, which leads to a decreasing internal field. As a result, the transient current decreases in magnitude. Second, at high photoinjection levels ($Q > CV$, where C is the sample capacitance and V is the applied field), the space charge may be formed by the drifting packet itself. In our case, this second type can be excluded because of the small value of injected charge ($Q \leq 0.1CV$).

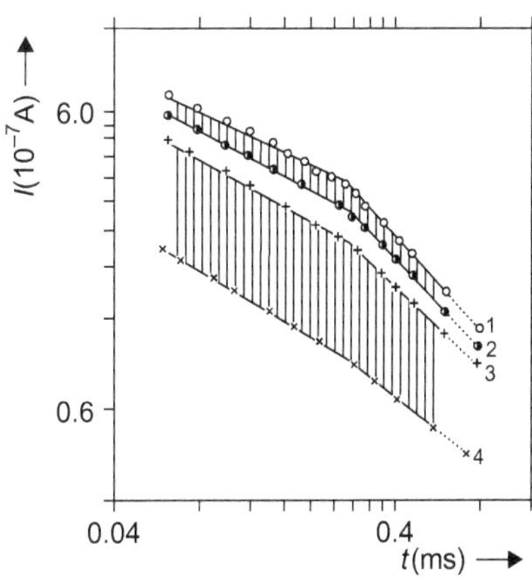

Figure 6.6 Transient hole current in a-As_2Se_3: (1) and (2) annealed sample, (3) and (4) photodarkened sample. The area between (1) and (2) and (3) and (4) indicates space charge accumulation during a 10-transit sequence with repetition frequency 10 Hz. $d = 3.4\,\mu m$, $E = 5.3\,V/\mu m$, $T = 293$ K.

Bulk-space-charge creation is shown in Fig. 6.6 in the areas between curves 1 and 2 and 3 and 4. The estimated amount of charge accumulated by deep traps during a 10-transit sequence increases 1.3–2 times at photodarkening and approaches about $10^{14}\,cm^{-3}$.

Consequently, shallow states that control charge transport and define the activation energy E_{μ}^{h} (about 0.4 eV for As_2Se_3) of the mobility should not undergo photo-induced changes. At the same time it should be stressed that the observed change in the current transient (namely, its increased dispersion and decreased magnitude) indicates photoinduced changes of deep states with $E_i > E_{\mu}^{h}$. Such behavior is probably related to enhanced carrier trapping by deep levels in photodarkened samples.

References

1. S.O. Kasap, in: A.S. Diamond, D.S. Weiss (Eds.), Handbook of Imaging Materials, second edn., Marcel Dekker, Inc., New York, 2002.
2. V.I. Mikla, J. Phys. Condens. Matter 8 (1996) 429.
3. N.F. Mott, E.A. Davis, in: Electronic Processes in Non-Crystalline Materials, second edn., Oxford University Press, Oxford, 1979.
4. K. Tanaka, in: Encyclopedia of Materials, Elsevier Science Ltd., Amsterdam, 2001, p. 1123.
5. W.E. Spear, J. Non-Cryst. Solids 1 (1969) 197.
6. J. Orenstein, M. Kastner, V. Vaninov, Phil. Mag. B 46 (1982) 23.
7. V.I. Mikla, Ph.D. Thesis, Odessa State University, Odessa, 1984.
8. V.I. Mikla, D.G. Semak, A.V. Mateleshko, A.A. Baganich, Phys. Status Solidi (a) 117 (1990) 241.
9. M. Abkowitz, R.C. Enck, Phys. Rev. B 25 (1982) 2567.
10. A.E. Owen, W.E. Spear, Phys. Chem. Glasses 17 (1976) 174.

7 Spectroscopic Studies of Gap States and Laser-Induced Structural Transformations in Se-Based As-Free Amorphous Semiconductors

Contents

7.1 Preparation of Amorphous Films: Essential Results and Interpretation

Amorphous Sb_xSe_{1-x} films were prepared by thermal evaporation in a vacuum at 10^{-6} Torr, and their electrical properties were examined. A combination of xerographic and TOF techniques was employed to study the gap states. Xerographic dark discharge experiments on a-Sb_xSe_{1-x} films indicated that the decay of the surface potential over the timescale of observation ($\sim$600 s) is essentially due to bulk thermal generation of electrons, and their subsequent sweep-out and depletion. Electron emission occurs from mid-gap localized states. When a-Se is alloyed with antimony, the dark discharge becomes more rapid due to an increase in the volume density of the mid-gap electron emission centers with antimony concentration. The repetition of the xerographic cycle over many iterations leads to the saturation of the surface residual voltage, which was used to determine the concentration N_t of deep electron traps. These cycled-up xerographic residual voltage measurements indicate that the saturated residual voltage increases with the addition of Sb due to an increase in the concentration of electron deep traps ($N_t \sim 10^{15}\,cm^{-3}$). The xerographic photosensitivity for Sb_xSe_{1-x} alloys is greater at longer wavelengths ($\lambda > 670$ nm) than for pure Se. TOF experiments indicate that electron transport in a-Sb_xSe_{1-x} alloys is controlled by a set of shallow traps at approximately 0.33 eV below E_c, whose concentration increases with Sb addition. Laser-induced transitory changes in the optical properties

Trap Level Spectroscopy in Amorphous Semiconductors. DOI: 10.1016/B978-0-12-384715-7.00007-3 and a phase transformation to crystalline modification (photocrystallization) were considered. The photoeffects observed critically depend on exposure (intensity) and exhibit threshold behavior. When the irradiation intensity is more than the threshold value, photoinduced crystallization takes place. The origin of this phenomenon is discussed on the basis of a microcrystalline model.

The physical properties of chalcogenide glasses that are arsenic free have attracted much interest for physical and technical reasons. Among these noncrystalline materials, the Sb_xSe_{1-x} glasses are attractive candidates for applications requiring low melting temperatures, low thermal conductance, and high viscosity. In particular, thin films of a Sb–Se system, which can undergo amorphous-crystalline phase transition, were studied as reversible optical recording candidates.

Chalcogenide glassy semiconductors, which are based on sulfides, selenides, and tellurides of the Main Group III–V elements, have recently gained much interest as materials for infrared-transparent optical fibers, photoconductors, photoresistors, optical memories, optoelectronic circuits, and so forth [1–4]. In general, glasses are often preferred over crystalline compounds with similar characteristics because of favorable mechanical and interfacing properties (i.e., properties that can be influenced by the presence of interfaces between various layers in multilayer structures and heterojunctions) and good processability. Further, a lack of translational regularity often makes it possible to tailor physical properties to specific applications by adjusting chemical composition. Selenium in the noncrystalline phase has a special attraction in semiconductor physics research not only because of its commercial importance as a xerographic photoreceptor material, but also because of its interesting physical properties [2–4]. The elemental nature, relative structural simplicity, ambipolarity, high photosensitivity, and so forth make Se a particularly suitable object for the study of impurities and/or alloying effects on the electrical properties. The alloying effects in a-Se have been well documented and reviewed for such additives as As, S, Te, P, e.g., Refs. [3, 5–8]. At the same time, there is only limited information about the basic properties of the amorphous Sb–Se system [9–11]. Recently, thin films of these materials have become of interest for use in data storage [10]. In the following sections, we report the effect of antimony on the deep and shallow states of selenium. These states are the thermal emission (trapping) centers, which control the slow dark decay, and the first and cycled-up residual surface potential on capacitively charged specimen films.

Glassy Sb_xSe_{1-x} alloys ($x < 0.05$) were prepared by conventional melt quenching. Cleaned silica tubes containing a mixture of the appropriate amount of constituents Sb and Se were evacuated to 10^{-5} Torr and sealed. The contents of the tubes were melted in a furnace and continuously agitated for 10 h to ensure good homogeneity. The melt was rapidly quenched in cold water from 800 K, and the cooling rate was estimated to be 200 K/s.

Amorphous films were prepared by thermal evaporation of the Sb_xSe_{1-x} source material from an open stainless steel boat onto preoxidized Al substrates in a vacuum of approximately 10^{-6} Torr. The substrate was cleaned in neutral detergent, deionized water, ethyl alcohol, and acetone and then oxidized in air at 1800°C before deposition. Deposition of the Sb_xSe_{1-x} alloys was performed at a substrate temperature of 300 K. It should be noted that while the substrate is aluminum, the layer directly under the film is aluminum oxide. Care was taken to avoid Sb fractionation. It should also be noted that when excluding the initial and final stages of the evaporation process by shuttering, the deposited antimony content is relatively uniform across the film thickness, as can be seen from our electron probe microanalysis data, and is close (±0.5 at.%) to the starting source value. A typical coating rate was 1–2 μm/min. The film thicknesses ranged from 10 to 20 μm. Prior to measurements, the Sb_xSe_{1-x} films prepared were aged in the dark under normal laboratory conditions (i.e., $T \approx 293$ K and relative humidity 70–80%) for 2–3 weeks to allow their physical properties to reach equilibrium.

Sb alloying effects on thermal generation and both deep and shallow trapping of carriers have been examined using xerographic and TOF techniques. In the xerographic measurement, the sample is charged to a potential V_0 by passing it under corotron (corona-charging) device that deposits charges of appropriate sign on the surface of the film. The surface potential is then measured by a transparent probe and an electrostatic voltmeter. The photoreceptor, the corona-charging units, and a transparent probe were housed in a well-shielded and dark environment. Following the initial charging process, the sample is exposed to strongly absorbed 450 nm step illumination from a tungsten light source, during which time the decay of the surface electrostatic potential is monitored by the electrostatic voltmeter. The surface potential decays to a potential V_r (termed the residual potential), and the resulting photoinduced discharge curve (PIDC) can be used to determine the xerographic photosensitivity of the sample. The preceding xerographic step can be repeated any number of times to obtain a cycled-up residual surface potential V_{rn} as a function of the xerographic cycle n.

The xerographic measurements have been complemented by TOF transient photoconductivity measurements to study the charge transport properties of Sb_xSe_{1-x} alloys. The experimental TOF technique has been described in the literature by a number of authors [3, 12, 13]. In a-Se-based materials, it is more useful to measure time-resolved transits in the current mode of operation because of the dispersion and relatively long transit at low fields. In the current case, we used a nitrogen gas laser to photoexcite the carriers. The nitrogen gas laser provided a short light pulse of approximately 10 ns at a wavelength of 337 nm. A semitransparent gold electrode was sputtered onto the surfaces of the films as the top electrode. Because the photocarriers are generated at the sample surface due to the high absorption coefficient of Sb_xSe_{1-x} glasses at 337 nm (>10^4 cm^{-1}), the species of drifting carriers can be chosen by changing the polarity of applied voltage across the sample. The transient current was amplified and displayed on a single-event storage oscilloscope. The TOF measurement was carried out under the single-shot mode of operation, and the sample was short-circuited and stored (relaxed) in the dark between each measurement to allow any bulk space charge buildup to decay. Small-signal conditions were maintained throughout all the measurements.

7.2 The Basic Properties

First, we begin with a brief review of some fundamental physical properties. We defined the glass transition temperature in the usual way—as the midpoint of the transition endotherm. Starting with 1 at.% Sb, our data for T_g show the monotonic increase with increasing Sb concentration. At the same time, we observe a change of slope (step-like behavior) between $x = 0$ and $x = 0.02$, signaling a change in the structure. For each composition, we carried out several experiments and all showed good reproducibility. Note that the experimental points represent average values with the size of the plotted bars comparable to the size of the plotted experimental point. It should be noticed further that microhardness and density show similar compositional trends with local rise around the same composition.

For all Sb–Se compositions used in this study, the dark DC conductivity can be expressed in the temperature range considered by an Arrhenius-type relation

$$\sigma = \sigma_0 \exp(-E_\sigma/kT)$$

where E_σ is the activation energy and σ_0 is the conductivity preexponential factor. The activation energy and the conductivity at 300 K are presented in Table 7.1.

It can be seen that both the conductivity and the activation energy decrease with increasing Sb content.

In the glassy system Sb–Se, the optical absorption edge (the Urbah tail region) shifted to longer wavelengths and the slope became smaller as the percentage of Sb increased. This is a general trend over most of the composition range. Absorption above the fundamental edge follows the familiar Tauc law

$$\alpha h\nu = C(h\nu - E_T)^2$$

where $h\nu$ is a photon energy, E_T is the Tauc gap, and C is a constant indicating how steeply the absorption rises with energy. For most compositions, there was no deviation from the square law at the highest absorption value measured in this study; exceptions are Se and films with less than 1 at.% Sb. The latter follow a linear law

$$\alpha h\nu = C_1(h\nu - E_1)$$

Table 7.1 The DC Conductivity (300 K), the Corresponding Activation Energy, and the Tauc Gap of Sb_xSe_{1-x} Glasses

Antimony content (at.%)	σ_{ROOM} **($\Omega^{-1}cm^{-1}$)**	E_σ **(eV)**	**dV/dt (V/s)**	$t_{V_0}/2$ **(s)**	E_0 **(eV)**
0	1×10^{-14}	1.05	7.8×10^{-2}	267	2.04
2	1×10^{-13}	1.02	2.0×10^{-1}	81	2.02
5	8×10^{-11}	0.90	1.83	76	
8	9×10^{-10}	0.84			
10	4×10^{-10}	0.76			1.85
20	1×10^{-9}	0.72			1.62
40	2×10^{-8}	0.62			1.50

in the range 2.1–2.5 eV. Here, C_1 is a constant and E_1 is an extrapolated optical gap. The addition of less than 1 at.% Sb to pure Se is enough to cause a complete transition from the anomalous linear behavior to Tauc's law. The room-temperature values of E_T (averaged over several samples) for various Sb_xSe_{1-x} compositions are given in Table 7.1. It should be noted that the Tauc gap E_T is close to that of pure Se for small Sb concentrations (<1 at.% Sb), and then it decreases almost linearly up to nearly 40 at.% Sb.

7.3 Dark Discharge

Typical dark discharge characteristics for pure Se and Sb_xSe_{1-x} photoreceptors are shown in Fig. 7.1 for compositions noted in the figure. It is apparent that for pure a-Se, the decay of the surface potential is relatively slow. Comparison of the respective characteristics for a-Sb_xSe_{1-x} with the dark discharge behavior of pure a-Se shows clearly that alloying a-Se with antimony increases the dark-decay rate. The discharge rates in a-Sb_xSe_{1-x} were not constant but decreased with time.

There are several physical processes that can lead to the decay of the surface potential. The currently accepted model for the dark decay in a-Se-based films involves [3, 14, 15]:

a. surface generation and injection of trapped electrons and their consequent transport across the sample;
b. substrate injection;
c. bulk thermal generation of carriers of one sign and depletion.

With relatively thick films (L = 10–50 μm) and a good blocking contact between a-Se-based films and the preoxidized Al substrate, the latter phenomenon dominates.

In a series of experiments carried out on a composition series of glassy Sb_xSe_{1-x} alloys, it was found that the time-dependent dark-decay rate of the potential to which

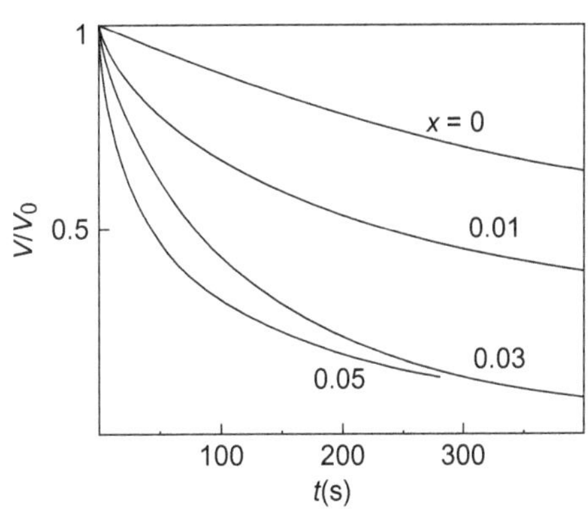

Figure 7.1 Dark discharge of surface potential on a-Sb_xSe_{1-x} layers.

a dark relaxed film has been charged is controlled by the depletion discharge process. In general, the xerographic depletion discharge model is based on bulk thermal generation involving the ionization of a deep mobility gap center to produce a mobile charge carrier of the same sign as the surface charge and an oppositely charged ionic center [14, 15]. Assuming negative charging, a mobile electron would be thermally generated, and the ionized center would be positive. As thermally generated holes are swept out by the electric field, a positive bulk space charge builds up with time in the specimen, causing the surface potential to decay with time.

Figure 7.2 depicts typical dark discharge data for four different compositions, which illustrate the predicted characteristics of depletion discharge behavior. Inflections in the log–log plots at the respective depletion times (marked by an arrow for the case of $Sb_{0.03}Se_{0.97}$ composition) are readily identifiable. From the temperature dependence of depletion time, it is estimated that the emitting sites are located approximately 0.9 ± 0.05 eV below the conduction band mobility edge. Some special problems can, however, complicate the observation of a depletion kink in pure a-Se. The dark discharge rate was typically so slow in a-Se that results were always perturbed by injection. The main reasons pure Se possesses good dark-decay characteristics are (a) the remarkably small number of integrated deep localized states ($\sim 10^{13}\,cm^{-3}$) in the mobility gap of a-Se and (b) the energy location of these states, which is deep ($E_t \sim 1.0$ eV) in the mobility gap so that the thermal generation process of carriers from these centers is slow [3, 6, 15].

It is found that in a-Sb_xSe_{1-x} alloys, electrons (the mobile carrier species) are depleted (n-type system) during dark decay, leaving behind a deeply trapped positive space charge. Note that the same situation prevails in alkali-doped a-Se [16].

7.4 Transient Photoconductivity

Let us first consider the carrier drift in pure amorphous selenium. Both the electron and hole drift mobility can be measured in a-Se by the TOF technique outlined earlier. At temperatures above 200 K, a well-defined transit pulse is observed. The transient

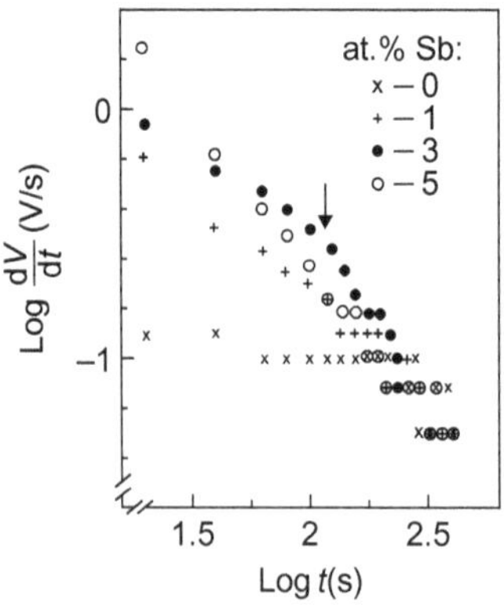

Figure 7.2 Typical dark discharge data for Sb–Se.

profile for a well-relaxed (dark-adapted) sample is a quasi-rectangular pulse. The photocurrent remains approximately constant up to 0.2 and 6 μs (for holes and electrons, respectively), then decreases abruptly. The signals indicate essentially Gaussian transport. The transit time t_T was defined to correspond to the break point in the photocurrent. This represents the transit time of the fastest carriers. The mobility, μ, is then calculated using $\mu = L^2/(t_T V_0)$, where V_0 is the applied voltage and L is the sample thickness. These room-temperature values of the drift mobility $\mu^h \approx 2 \times 10^{-1}\,cm^2/V/s$ and $\mu^e \approx 7 \times 10^{-3}\,cm^2/V/s$ are in remarkable good agreement with reported values [3, 5–7, 12, 13]. Note that for a-Sb_xSe_{1-x} films deposited on room-temperature substrate, it was not possible to detect any pulses associated with the transit of hole carriers. In all samples, hole response showed a rapid decay with no apparent break.

There are two plausible explanations. First, the signal is limited by the presence of deep gap states with extremely high efficiency for carrier trapping. Second, it might be argued that alloying may have increased the conductivity of the samples so that TOF experiments are no longer applicable. However, results showed that although the conductivity increases with the addition of Sb, it remains sufficiently low ($\leq 10^{-11}\,\Omega^{-1}cm^{-1}$) for TOF experiments to be applicable. It is interesting that hole response in As-alloyed a-Se has also been found to be undetectable in the range 2–4 at.% As [5, 17].

The effect of Sb on electron transport is not so drastic. Although Sb alloying increases the transit time dispersion, the transit time shown contains a clearly identifiably break in the waveform. The electron drift mobility in a-Sb_xSe_{1-x} alloys exhibits Arrhenius behavior. The experimentally observed activation energy of a-Se—namely, $E_\mu \approx 0.33 \pm 0.01$ eV—remains almost insensitive to the addition of Sb. At the same time, the mobility decreases with increasing Sb content. It is reasonable, within the framework of a shallow trap-controlled mobility model [18], to interpret our TOF observations in the following manner. In pure Se, electron transport is controlled by a narrow manifold of traps located about 0.33 eV from the conduction band mobility edge. The addition of Sb to Se, we believe, broadens the distribution of shallow traps, thus increasing the relative dispersion of photocurrent transients. The decrease in electron mobility with increasing Sb content can be accounted for by the increase in the density of shallow traps, with E_μ remaining constant at approximately 0.33 eV.

7.5 Photoinduced Discharge Characteristics

In essence, the xerographic photosensitivity (S) of a photoreceptor material determines the rate of decay, dV/dt, of the electrostatic surface potential of the samples during photoinduced discharge (PID). The xerographic photosensitivity definition adopted here is simply based on the amount of light energy required for the surface potential to decay to half of its original value ($V_0/2$) during PID (i.e., the fractional change in the surface potential per unit of light exposure).

Figure 7.3 shows a dark-decay curve and a PIDC for the a-$Sb_{0.03}Se_{0.97}$ film. It can be seen that the sample exhibits relatively little dark decay. Nevertheless, in order to

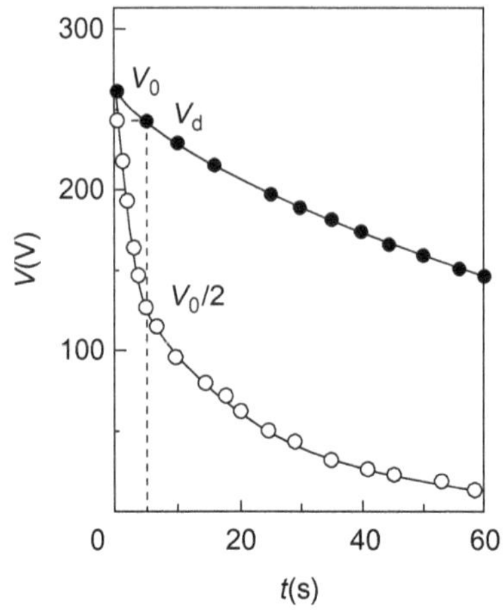

Figure 7.3 Dark-decay curve (solid circles) and photoinduced discharge curve (open circles) for $Sb_{0.03}Se_{0.97}$.

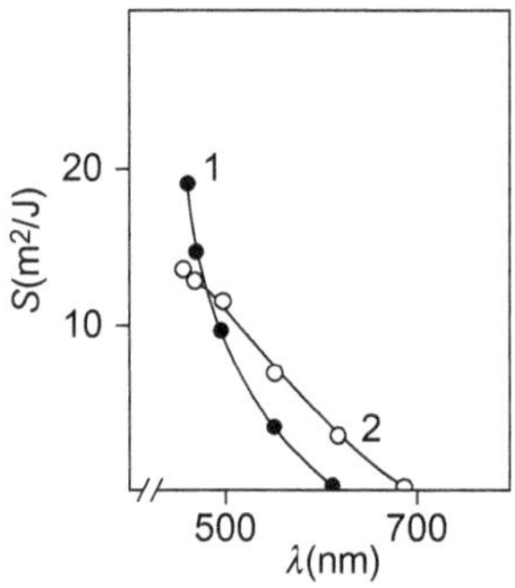

Figure 7.4 Xerographic photosensitivity (S) versus exposure wavelength (λ) for pure Se (1) and $Sb_{0.03}Se_{0.97}$ (2) alloy films.

take into account the surface charge reduction during illumination and to evaluate S accurately, the contribution of dark discharge to the total change in the surface potential during PID was subtracted ($V_0 - V_d$ in Fig. 7.3).

The xerographic spectral response for both a-$Sb_{0.03}Se_{0.97}$ and pure Se are shown in Fig. 7.4.

It is apparent that as Sb is added to a-Se, the photosensitivity at a particular wavelength increases. More precisely, this means that the xerographic photosensitivity for the Sb_xSe_{1-x} alloy is somewhat greater at longer wavelengths ($\lambda > 670$ nm) than for pure Se and smaller at shorter wavelengths ($\lambda < 500$ nm). Note that other compositions of Sb–Se alloy showed similar trends. The result of increased xerographic photosensitivity of the Sb–Se alloys at longer wavelengths suggests that the addition of antimony to a-Se causes a reduction of the band-gap of the material. It should be noticed here that the xerographic photosensitivity depends not only on the absorption

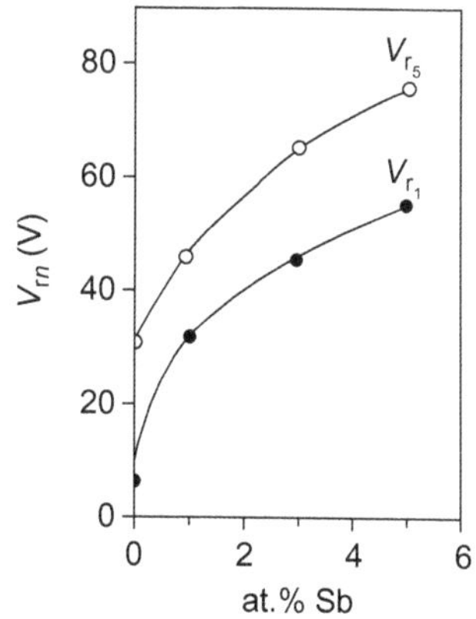

Figure 7.5 Effect of antimony on the residual potential (measured after first and fifth cycles) of Sb–Se alloy.

coefficient α, but also on the quantum efficiency η for generating mobile charge carriers as well as the transport properties (the $\mu\tau$ product) [3].

For the films under examination, the residual voltage V_r (a measurable surface potential at the end of the illumination) increases with the Sb content (Fig. 7.5).

The residual potential is due to trapped electrons in the bulk of the specimen. The simplest theoretical model, which is based on range limitation and weak trapping ($V_r \ll V_0$), relates V_r to $\mu\tau$ (the drift mobility μ and lifetime τ product) via the Warter equation [19]:

$$V_r = L^2/2\mu\tau$$

where L is the sample thickness. For example, addition of 3at.% Sb leads to a change in the first-cycle residual voltage from 4 to 44V, which is equivalent to a change of the carrier range $\mu\tau$ from 10^{-7} to 10^{-6}cm^2/V. Substituting $\mu^e \sim 7 \times 10^{-3}$cm^2/(V s) for pure Se and $\mu^e \sim 6 \times 10^{-4}$cm^2/(V s) for $Sb_{0.03}Se_{0.97}$ into the corresponding equation, we find carrier lifetimes $\tau \approx 2 \times 10^{-4}$s and $\tau \approx 1.3 \times 10^{-3}$s in a-Se and a-$Sb_{0.03}Se_{0.97}$. It is necessary to note here that in general, bulk deep-trapping lifetimes computed from the first-cycle residuals are in agreement with lifetimes measured in the TOF mode under range-limited conditions.

Figure 7.6 displays the buildup of the residual voltage V_{rn} on a-$Sb_{0.03}Se_{0.97}$ film with the number of xerographic cycles n. The rate of V_{rn} decreases with cycling. Then, for large n (≥6 in our case), V_{rn} tends to a saturation value V_{rs}.

As described earlier [20], the saturation residual potential provides an experimental measure of the integrated number of deep traps (trap-release rates are much slower than those from shallow traps which control drift mobility). V_{rs} is then simply given by

$$V_{rs} = eN_tL^2/2\varepsilon$$

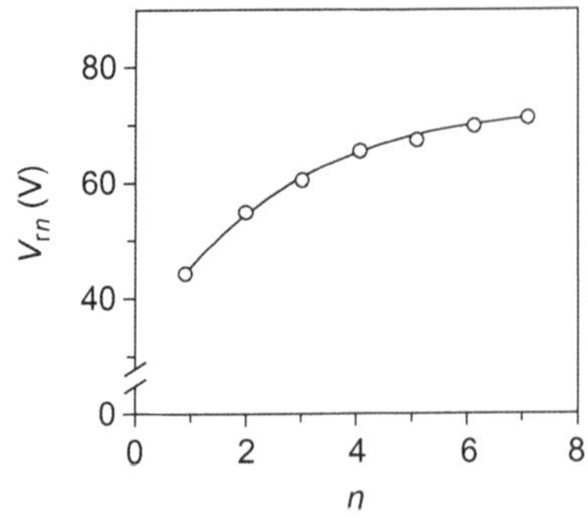

Figure 7.6 The buildup in the residual voltage with number of xerographic cycles in a-$Sb_{0.03}Se_{0.97}$ films.

where N_t is the deep-trap concentration and ε is the dielectric constant.

Both the first residual and the cycled-up saturated residual potential are sensitive to alloying. For example, when pure amorphous Se films are alloyed with antimony, the buildup of the residual potential occurs more rapidly toward a much higher saturated residual potential. We obtain, for instance, $N_t \sim 2 \times 10^{14} cm^{-3}$ and $N_t \sim 10^{15} cm^{-3}$ for a-Se and $Sb_{0.03}Se_{0.97}$, respectively.

The preceding photoelectric properties of a-Sb_xSe_{1-x} alloys can be at least qualitatively explained by using concepts based on charged structural defects, called valence alternation pairs (VAPs) or intimate valence alternation pairs (IVAPs). These correspond to some of the chalcogen atoms being under- and overcoordinated [21]. It seems reasonable that dark discharge and residual voltage buildup involve essentially the same species of localized centers. Further, it is also possible that we may be observing amphoteric behavior by IVAPs. For pure Se, an IVAP comprises over- and undercoordinated selenium atoms Se_3^+ and Se_1^- in close proximity. An IVAP center would be seen as a "neutral trap" by the carrier. The capture of an electron by Se_3^+ exposes the negative charge on Se_1^- and explains the residual voltage detected. At the same time, the emission of an electron from the Se_1^- uncovers the positive charge on Se_3^+, which causes the dark decay. Although the question remains whether the neutral center is a neutral dangling bond D0 type defector whether it is an valence alternation pairs (VAP) or intimate valence alternation pairs (IVAP) defect, the measured radius of 3 Å in Ref. [22] is representative of an IVAP capture radius.

There are a number of desirable electrical characteristics that a useful photoreceptor should exhibit. Our experimental results show that dark-decay rate, residual voltage, and drift mobility are all sensitive to antimony. Carrier drift mobility decreases with antimony, whereas dark-decay rate and residual voltage increase. All these effects are undesirable in xerography. At the same time, the main advantage of a-Sb_xSe_{1-x} is that its spectral response can be readily shifted to longer wavelengths by increasing the Sb content, which allows for photoreceptor designs that can respond to a variety of illumination spectra. In addition, it may be possible to improve the charge transport parameters by halogen doping. Indeed, Cl is known to compensate for As-induced (same group as Sb) deep traps in a-Se [23]. By using a double-layer photoreceptor consisting of a thin a-Sb_xSe_{1-x} layer for photogeneration and a thick a-Se layer for charge transport, the xerographic parameters can be further improved.

Thus, the mobile carrier species controlling the xerographic depletion discharge in a-Sb–Se alloys are electrons. Thermal generation of free electrons in a-Sb_xSe_{1-x} is accompanied by the simultaneous formation of a deeply trapped positive space charge. It was shown that antimony alloying progressively enhances the free electron thermal generation rate relative to the pure specimen.

As apparent from the large xerographic residual potentials for Sb_xSe_{1-x} alloys, the addition of Sb to a-Se seems to greatly increase the concentration of deep localized states within the mobility gap of the material.

The results indicate that in the long-wavelength region (e.g., $\lambda \sim 600\,\text{nm}$), the photosensitivity for the a-Sb_xSe_{1-x} films is higher than for the pure selenium, probably due to a greater quantum efficiency.

In such an exotic field of materials science as the amorphous (disordered) solids, one of the fundamental problems studied extensively is how to obtain insight into the structure. Currently, it seems that versatile studies are needed to elucidate the amorphous structure. In other words, in addition to various direct and indirect structural techniques performed under fixed conditions (X-ray diffraction, Raman scattering, infrared absorption, X-ray absorption, to name a few), the investigation of structural modifications introduced by changes in composition, temperature, or pressure or induced by band-gap illumination may prove fruitful.

One of the properties of a class of materials known as chalcogenide glasses is that they exhibit a wide spectrum of photoinduced effects. Photoinduced phenomena have recently been extensively studied (see corresponding references in previous sections), partly as an interesting subject for fundamental research in the field of disordered solids and partly due to the potential application of these phenomena in opto(photo)electronics (xerography and xeroradiography, optical memories, optical circuits, photoresists, etc.).

Among these phenomena, so-called reversible photodarkening and photocrystallization are the most interesting. The changes in various physical and chemical properties of chalcogenide glasses under band-gap illumination have been detected. Since 1968, researchers have known that band-gap illumination of amorphous selenium films increases the growth rate of crystallites [24]. This phenomenon was later utilized by others to develop images on selenium layers 100–150 μm thickness on gold. Although several mechanisms of photodarkening and photocrystallization have been proposed, the details of these apparently simple phenomena remain ambiguous.

In this section, we will examine some features of these two phenomena—room-temperature reversible photodarkening and photocrystallization—in amorphous semiconductors films of Sb_xSe_{1-x}. For the starting material, pure amorphous Se was chosen. One of the elemental amorphous materials, Se may be extremely suitable for discussing essential features and relationship (if such exists) between the two phenomena chosen for discussion. In addition, the effect of small amounts of antimony (a few percent) on photodarkening and photocrystallization of a-Se is especially interesting—not only from the point of view of compositional disordering, but also because of desirable recording properties and peculiarities of electronic transport for amorphous Sb_xSe_{1-x} films [25].

Samples investigated were 0.3–3.0 μm amorphous Sb_xSe_{1-x} ($0 \leq x \leq 0.05$) films. These were preferentially prepared by conventional vacuum evaporation onto room-temperature silica-glass substrates (plates). The a-Sb_xSe_{1-x} source material was made by the usual melt-quenching technique. Cooling rate was estimated to be 100–200 K/s. Prior to measurements, the films prepared were aged at laboratory conditions (natural aging) for several weeks to allow their structure to equilibrate.

Three kinds of measurements were performed:

a. Transmission photodarkening experiments, where the samples were illuminated at near-normal incidence by a helium–neon laser operating at 633 nm. The transmission of the samples was probed using a portion of relatively low (~3 mW) intensity from the He–Ne laser output and detected by a photomultiplier. In this experiment, inducing and probing light propagate in parallel. The transmission was measured as a function of exposure time and intensity as well as the sample composition.

b. Holographic experiments. The current grating technique is the conventional method: a grating is produced by two interfering beams intersecting at a sample surface. The measured parameter η, the diffraction efficiency, is the ratio of the corresponding intensities of the probe beam I_0 and the first-order diffracted one I.

c. Structural probes. Right-angle Raman spectra and X-ray diffraction were measured at room temperature.

Photodarkening and photocrystallization of a-Sb_xSe_{1-x} films were induced by linearly polarized light with a wavelength of 633 nm emitted from a He–Ne laser. The light intensity was varied at $0.5\text{–}2.5 \times 10^2\,\mathrm{W/cm^2}$.

7.6 Optical Properties

Figure 7.7 shows the change in transmissivity as a function of exposure time for amorphous Sb_xSe_{1-x} films. Here, $T_{rel} = T_{ir}/T_{un}$ denotes the relative transmissivity and T_{ir} and T_{un} are the irradiated and unirradiated samples, respectively. One can clearly

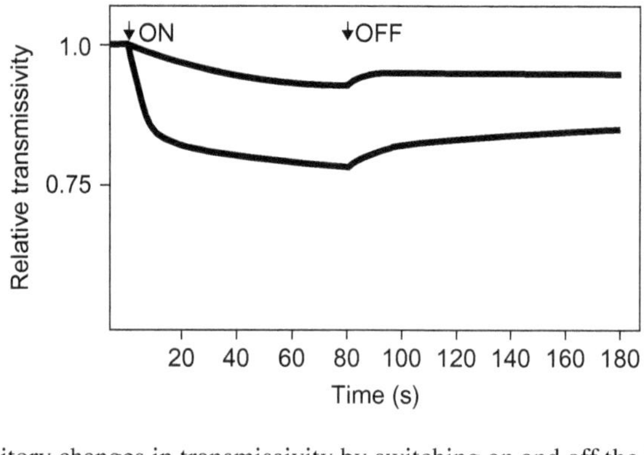

Figure 7.7 Transitory changes in transmissivity by switching on and off the irradiation of 633 nm in a-Sb_xSe_{1-x} films 1 and 5 at.%. Thickness is 0.8 μm and intensity 0.5 W/cm^2. Sb concentration is 1 and 4 at.% (curves 1 and 2, respectively). On- and off-periods of illumination are shown by arrows.

see a decrease in T_{rel} with the irradiation time. Light-induced change of transmissivity is transient: the initial value of T_{rel} is restored after switching off the light. This is not surprising because the glass transition temperature of compositions examined is approximately room temperature (pure a-Se) or slightly above this value (Se containing <5 at.% of Sb additives). Note that light with low intensity ($I < 10\,\mathrm{mW/cm^2}$) could not induce appreciable permanent effects, despite long exposure times. Only transitory behavior in T_{rel} is observed.

On the other hand, with increasing exposure (more precisely, its intensity), significant irreversible changes in transmissivity are observed (Fig. 7.8). We see that the characteristic at $I < I_{th}$ and $I > I_{th}$, where I_{th} is the threshold intensity, are quite different. In general, the T_{rel} versus E curve (E denotes exposure magnitude) may be divided into three parts:

a. initial, with a rapidly varying response—transient transmission change;
b. middle, slowly varying portion of the darkening curve; sometimes this part is marked by the beginning of a plateau region for intensities $I \approx I_{th}$;
c. final, with a significant (up to 0.8–0.9) decrease in T_{rel}—permanent, irreversible change.

The photoeffect is completed by a saturation region; as for cases (a), (b), and (c), the onset of the latter is intensity dependent.

Changes in transmissivity for $I \geq I_{th}$ (cases (b) and (c), respectively) may be attributed to crystallization transformation (discussed later).

Nearly the same behavior can be discerned for holographic recording properties. The η versus E data also exhibit several stages, depending on energy density. A typical response of light intensities diffracted from gratings formed on films is shown in Fig. 7.9 (low intensities) and Fig. 7.10 (high intensities): an initial increase to $\eta_{max} = 0.012\%$, a slight decrease, then it monotonically increases, with E showing the maximum at $\eta = 6\%$. The high-energy density side of the preceding maximum is probably caused by the optical absorption increase for the probing beam due to photocrystallization.

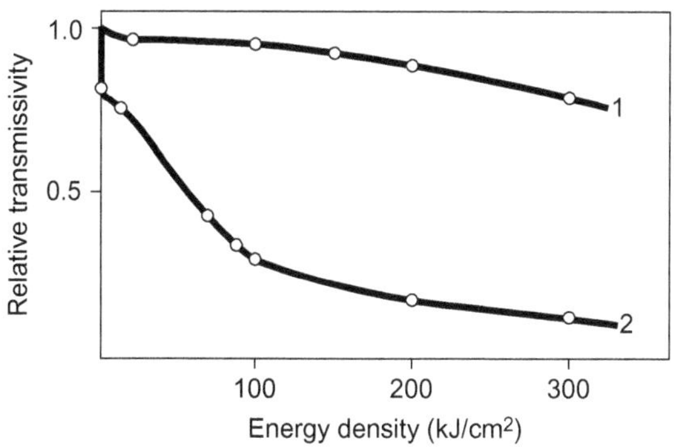

Figure 7.8 The relative transmissivity versus energy density in amorphous Sb_xSe_{1-x} films exposed to intense laser illumination at 633 nm. Thickness is 0.8 μm and intensity $2.5 \times 10^2\,\mathrm{W/cm^2}$. Sb concentration for curves 1 and 2 are 1 and 5 at.%, respectively.

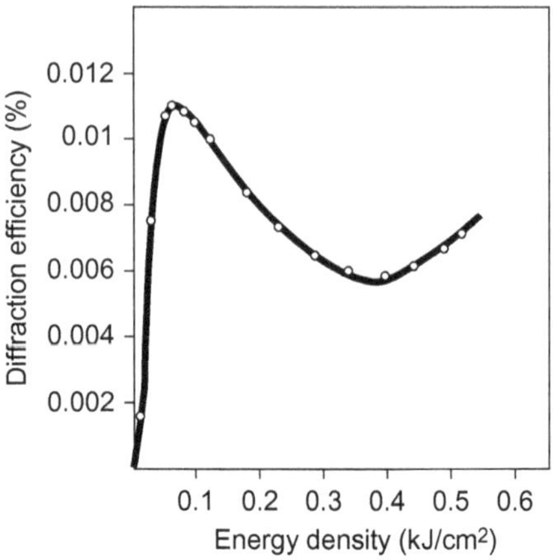

Figure 7.9 Diffraction efficiency versus energy density for pure amorphous selenium in the low-energy-density region. The lines are eye-guides.

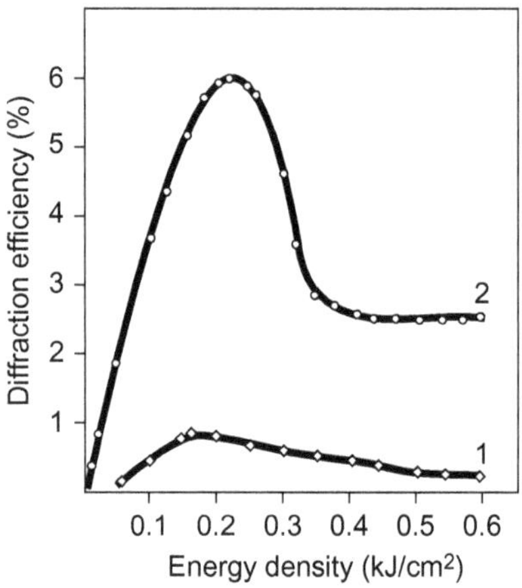

Figure 7.10 A change of diffraction efficiency in a-Sb_xSe_{1-x} samples induced by laser irradiation. $x = 0$ and 0.03 in curves 1 and 2, respectively. Thickness is 1.25 μm and $I = 1.25\,W/cm^2$.

The degree of change in T_{rel} and η increases with antimony content (see the corresponding figures). Apart from this, the addition of Sb to a-Se shifts the crystallization onset to higher exposure values.

7.7 Structural Transformation

Figure 7.11 shows the evolution of Raman spectra by increasing the light intensity of illumination. It is important to point out the existence of a certain threshold intensity of the incident laser beam—below I_{th}, the initial shape of the spectrum was recovered after turning off the illumination; above it, the spectrum further transforms

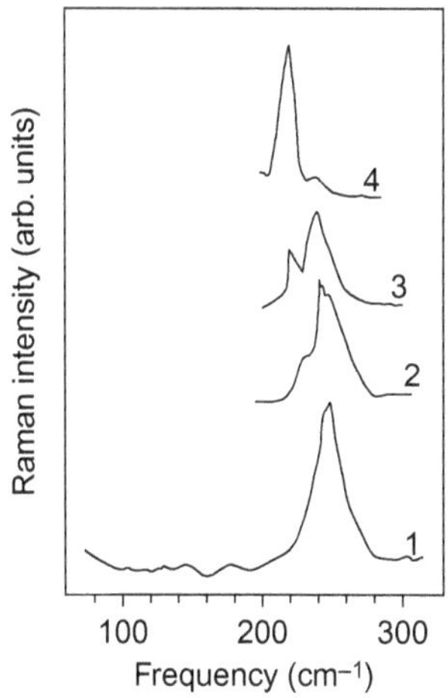

Figure 7.11 Transformation of Raman spectra in amorphous $Sb_{0.03}Se_{0.97}$, subjected to exposure of linearly polarized light: curve 1, reference Raman spectrum of amorphous state; curves 2–4, after exposure to $E = 3$, 5, and 6 kJ/cm^2, respectively. The inducing intensity was $I = 1.25$ W/cm^2.

(spontaneously), even in the absence of excitation. Note that the samples illuminated at $E < E_{th}$ are amorphous, while the samples illuminated at $E > E_{th}$ exhibit well-pronounced crystalline features in Raman spectra.

Some comments may be needed here for the Raman spectra. First, the spectra of the amorphous Sb_xSe_{1-x} samples before irradiation are close to those for pure selenium. The only additional feature observed is the appearance of the approximately 190 cm^{-1} band of $Sb_2Se_{3/2}$ structural units. Second, the exposure values at which the spectra start to transform vary with the addition of Sb. Third, under intense irradiation, Raman spectra show a gradual intensity redistribution for the peaks at 237 and 255 cm^{-1}. When the radiation power is further increased, the intensity of 237 cm^{-1} peak grows continuously with a simultaneous decrease of the 255 cm^{-1} spectral feature. Finally, it appears that Sb addition in such a quantity (<5 at.%) has no appreciable influence on the photocrystallization product. Actually, only the 237 cm^{-1} Raman band of hexagonal Se contributes to the spectra of photocrystallized Sb_xSe_{1-x} films. Thus, it is indicated clearly that on introducing a small quantity of additives to selenium glass, there is no appreciable influence on the crystallization product. This can be explained in an unambiguous way: the small quantity of antimony is clustered in the glass matrix. Further, in the case when the sample was illuminated, the Se crystallized out, while the additives still remained in the disordered (glassy) state. Even though a higher or lower E_{th} is found in samples of different chemical composition, there is little influence on the kind of crystallized product. In addition, it should be noted that an evolution in the shape of Raman spectra similar (at least qualitatively) to that presented here has also been observed for amorphous As_xSe_{1-x} alloys over the concentration range 0–10 at.% As.

X-ray measurements also indicate photocrystallization at exposure values $E > E_{th}$. Typical results, i.e., X-ray diffraction patterns of photocrystallized samples (this stage refers to curve 5 of Fig. 7.11), show four crystalline peaks located at $2\theta = 24°$, 30°, 41°, and 45°. These can be indexed as 100, 101, 110, and 111 of the hexagonal Se crystal, respectively. Similar X-ray diffraction patterns were obtained by Ishida and Tanaka [26] examining photoinduced birefringence in a-Se thin films.

The current experimental results distinguish three successive stages of laser-induced changes in a-Sb_xSe_{1-x} with regard to the irradiation energy density:

I. The initial stage presents the dynamic (transient) photoinduced effects in T_{rel}, η, and so forth. Selenium as elemental chalcogen exhibits reversible photodarkening. However, the phenomena have not been basically studied. The reason for this is that permanent (quasi-stable) photoinduced changes can only be realized at low ($T < 200$ K) temperatures—a complete recovery of initial parameters was observed for samples annealed at room temperature. It is essential to note that the sample remains in the *amorphous* state before, during, or even after irradiation. Reasonably, one may argue that the observed dynamic (transitory) change in transmissivity under light irradiation may be caused by thermal excitation (in other words, they are thermal in origin). Nevertheless, due to the following arguments, we do not believe this is a dominant factor, although some heat-up of the sample cannot be fully eliminated.

- **a.** The preceding photoeffect shows a dependence on the substrate material (note that we examined several types of the latter—polymethylmetacrylat, quartz glass, mica foils, etc.). It is evident that for the substrate materials mentioned, the heat dissipation conditions are essentially different.
- **b.** The lack of any noticeable variation in T_{rel} (or η) behavior for samples of different film thickness significantly reduces the possibility of the effect being due to changes in temperature during illumination.

  Some photoelectronic properties exhibit similar dynamic photoinduced changes [27]. We relate the transient changes in the transmissivity (photodarkening) to changes in metastable deep states in the mobility gap. The states are associated with charged structural defects (e.g., VAPs, IVAPs), which correspond to some of the chalcogen atoms, over- and undercoordinated. For pure Se, an IVAP comprises Se_3^+ and Se_1^- in close proximity. Band-gap illumination initiates conversion of such states into ones with greater density or capture cross section.

II. and **III.** The intermediate and final stages are both associated with crystallization: the former with microcrystallite formation, the latter with microcrystallite enlargement and its concentration (nucleation and growth of crystalline structures).

In order to discuss an essential feature of the laser-induced structural change, we must first consider the structure of amorphous films. Almost all available structural data indicate that the main constituent of a-Sb_xSe_{1-x} ($x < 0.05$) is the chain molecule, although molecules with Sb branching sites may be contained in minority. That is, we assume that the local structure of a-Se containing Sb additives resembles the hexagonal (trigonal) Se structure. Accordingly, in atomic structural terms, a-Sb_xSe_{1-x} with $x < 5$ at.% may be characterized as a quasi-one-dimensional chain structure.

Plausible explanations for the current experimental observations can provide the quasi-crystalline model proposed by Zhdanov et al. [28], Shimakawa et al. [29],

Tanaka and Ishida [30], and Fritzsche [31]. Following Fritzsche's phenomenological idea, we assume that the amorphous structure of annealed and then dark-rested films contains some anisotropic elements. When aligned in random orientation, the structure appears to be isotropic—a characteristic inherent to amorphous solid-state materials. If the preceding amorphous samples are exposed to linearly polarized light, structural elements (chain segments) tend to align in some preferential orientation that is nearly perpendicular to the electric field vector of linearly polarized light. Because Se-rich alloys in their noncrystalline form can be regarded as a kind of a "soft" material (its atomic structure is flexible due to a twofold coordination of chalcogen atoms), chain segment orientation can be influenced relatively easily. The intensity difference between the horizontal and vertical configurations of the 101 peak (observed in a series of X-ray diffraction patterns) seems to be consistent with the preferential alignment of the chain molecules perpendicular to the electric field vector of linearly polarized light. Accordingly, in spite of this microcrystalline model, if the crystallization process proceed in the manner described earlier, and if the structural elements mentioned are responsible for the photoinduced anisotropy in amorphous state, we may speculate that such an element with a quasi-crystalline orientation induced in the amorphous phase by linearly polarized illumination of low intensity (referred to as the *initial* stage) becomes a nucleus that grows as oriented crystals. The energy density needed for the latter process should be greater than the threshold energy E_{th}. Although microscopic structures giving rise to the light-induced transitory changes and photocrystallization still remain controversial, it is assumed that the latter phenomenon arise from changes at more extended than atomic scale structures.

Finally, it is interesting to consider the preceding results in relation to other Se-based binary alloys containing as constituent group V elements. For example, the same behavior is observed for the As_xSe_{1-x} system when the latter was subjected to band-gap illumination. Moreover, we find that many of the physical properties (glass transition temperature T_g, density d, band-gap energy E_0, etc.) of the Sb_xSe_{1-x} system change around $x = 0.01$–0.02. The same behavior is observed for the As_xSe_{1-x}, which has local extrema (inflections) in several properties below the usual threshold that is near the composition $x \approx 0.04$. A possible explanation suggested is that the peculiarities mentioned are associated with the topological threshold that occurs when the chain-ring-like structure changes to a chain-like structure [32, 33].

Very important is the observed similarity between the light-induced changes in pure amorphous Se and Se containing Sb or As additives. The similarity is possibly the clue to a more general understanding of various photoinduced changes, photocrystallization, and compositional dependency in chalcogenide vitreous semiconductors. Further experiments would be very helpful in this context.

Based on the preceding results, we can say that linearly polarized light from the band-gap absorption region can induce either transitory changes or crystallization transformation in amorphous Sb_xSe_{1-x} alloys. These two phenomena critically depend on exposure, show threshold behavior, and seem to arise from apparently different mechanisms: defect states or some kind of structural units given preferential orientation under the action of linearly polarized light.

References

1. A. Madan, M.P. Shaw, The Physics and Applications of Amorphous Semiconductors, Academic Press, Boston, MA, 1988.
2. K. Tanaka, Curr. Opin. Solid State Mater. Sci. 1 (1996) 567.
3. S.O. Kasap, in: A.S. Diamond (Ed.), Handbook of Imaging Materials, Marcel Dekker, Inc., New York, 1991, p. 329.
4. J. Rowlands, S. Kasap, Physics Today 11 (1997) 24.
5. J. Schottmiller, M. Tabak, G. Lucovsky, A.I. Ward, J. Non-Cryst. Solids 4 (1970) 80.
6. M. Abkowitz, J.M. Markovich, Solid State Commun. 44 (1982) 1431.
7. N.F. Mott, A.E. Davis, Electron. Processes Non-Cryst. Mater., Clarendon Press, Oxford, 1979.
8. C. Wood, R. Mueller, L.R. Gilbert, J. Non-Cryst. Solids 12 (1973) 295.
9. M.H. El-Zaidia, A. El-Shafi, A.A. Ammar, M. Abo-Ghozala, Thermochim. Acta 116 (1987) 35.
10. F.V. Pirogov, J. Non-Cryst. Solids 114 (1989) 76.
11. V.I. Mikla, Phys. Status Solidi (b) 182 (1994) 325.
12. W.E. Spear, J. Non-Cryst. Solids 1 (1969) 197.
13. G. Pfister, H. Scher, Adv. Phys. 27 (1978) 747.
14. A.R. Melnyk, J. Non-Cryst. Solids 35/36 (1980) 837.
15. M. Abkowitz, F. Jansen, A.R. Melnyk, Phil. Mag. B 51 (1985) 405.
16. M. Abkowitz, F. Jansen, J. Non-Cryst. Solids 59/60 (1983) 953.
17. V.I. Mikla, Mater. Sci. Eng. B 49 (1997) 123.
18. J.M. Marshall, F.D. Fisher, A.E. Owen, Phys. Status Solidi (a) 25 (1974) 419.
19. P.J. Warter, Appl. Opt. Suppl. 3 (1985) 65.
20. M. Abkowitz, R.C. Enck, Phys. Rev. B 27 (1983) 7402.
21. M. Kastner, D. Adler, H. Fritzsche, Phys. Rev. Lett. 37 (1976) 1504.
22. B. Polischuk, S.O. Kasap, V. Aiyah, M. Abkowitz, Can. J. Phys. 69 (1991) 364.
23. S.M. Vaezi-Nejad, C. Juhasz, J. Mater. Sci. 23 (1988) 3286.
24. J. Dresner, G.B. Stringfellow, J. Phys. Chem. Solids 29 (1968) 303.
25. F.V. Pirogov, J. Non-Cryst. Solids 114 (1989) 76.
26. K. Ishida, K. Tanaka, Phys. Rev. B 56 (1978) 206.
27. V.I. Mikla, I.P. Mikhalko, V.V. Mikla, Mater. Sci. Eng. B 64 (1999) 1.
28. V.G. Zhgdanov, V.K. Malinovsky, Phys. Status Solidi (a) 52 (1979).
29. K. Shimakawa, A. Kolobov, S.R. Elliott, Adv. Phys. 44 (1995) 475.
30. K. Tanaka, K. Ishida, J. Non-Cryst. Solids 227 (1998) 673.
31. H. Fritzsche, Phys. Rev. B 52 (1995) 15854.
32. V.I. Mikla, J. Phys. Condens. Matter. 9 (1998) 9209.
33. S.O. Kasap, D. Tonchev, T. Wagner, J. Mater. Sci. Lett. 17 (1998) 1809.

www.ingramcontent.com/pod-product-compliance
Ingram Content Group UK Ltd.
Pitfield, Milton Keynes, MK11 3LW, UK
UKHW020124290926
15471UKWH00003B/266

9 780323 165037